Rhetorical Touch

Studies in Rhetoric/Communication
Thomas W. Benson, Series Editor

Rhetorical Touch

DISABILITY, IDENTIFICATION, HAPTICS

Shannon Walters

The University of South Carolina Press

Published by the University of South Carolina Press
Columbia, South Carolina 29208

www.sc.edu/uscpress

Manufactured in the United States of America

23 22 21 20 19 18 17 16 15 14 10 9 8 7 6 5 4 3 2 1

Library of Congress Cataloging-in-Publication Data

Walters, Shannon, author.
Rhetorical touch : disability, identification, haptics / Shannon Walters.
pages cm. — (Studies in rhetoric/communication)
Includes bibliographical references and index.
ISBN 978-1-61117-383-3 (hardback) — ISBN 978-1-61117-384-0 (ebook) 1. Rhetoric.
2. Touch—Psychological aspects. 3. People with disabilities. 4. Haptic devices. I. Title.
P301.W355 2014
808—dc23

2014007287

This book was printed on recycled paper with 30 percent postconsumer waste content.

For my family—past, present, future

Contents

Illustrations

Series Editor's Preface

In *Rhetorical Touch: Disability, Identification, Haptics,* Shannon Walters explores the long history of touch as a topic and as a figure in rhetorical theory, starting with the fifth-century B.C.E. sophist and teacher of rhetoric Empedocles, who taught Gorgias, who in turn debates Socrates in Plato's dialogue *Gorgias.* Touch reappears through the rhetorical theorizing of Aristotle and Kenneth Burke. Touch, argues Walters, is a neglected sense in rhetorical theory that on closer inspection may be seen to infuse the language and conceptual structure of rhetoric. At the same time, Walters shows how touch for a person experiencing disability, physical or neurological, informs the life world of the disabled person, and at the same time how touch becomes itself rhetorical, a resource for identification with others, and a way of knowing, feeling, and communicating—both a limit and a resource. Walters proposes a theoretical understanding that relates the elements of traditional rhetoric—*ethos, pathos,* and *logos*—to the tactile rhetoric of sophistic theory—felt *logos, mētic-ethos* (embodied intelligence), and *kairotic-pathos,* as an appeal by bodies in close contact. This conception, Walters shows, has the merit of being able to explain and facilitate the communication of the disabled and of the temporarily abled. Walters explores in detail the way touch is used as a rhetoric and as a theme of rhetoric by disability advocates, and by such cultural figures as Helen Keller and Temple Grandin, among others.

Thomas W. Benson

Acknowledgments

This book is the work of many hands. Through the years over which it has taken shape, I have had the pleasure of learning from and working with a wide range of invaluable colleagues, teachers, students, friends, collaborators, and other supporters. My attempts at acknowledgment will surely only scratch the surface of this deep source of intellectual energy, but I will try.

I have deep appreciation for the rich and interdisciplinary network of scholars and students at Pennsylvania State University, where some of my first stabs at the questions of this book were made. Susan Squier taught me how to ask crucial questions, to recognize the stakes of important issues, and to enjoy the complexities of interdisciplinary engagement. She has been a valuable role model to me, radiating energy and enthusiasm as scholar, teacher, community member, and mentor. Jack Selzer supported this project at crucial stages, teaching me to see the bigger picture surrounding my questions and encouraging me to see their potential as a part of a larger conversation. Stuart Selber urged me to consider the lively connections among disability, accessibility, and technology that I have explored in this project and in others. Michael Bérubé encouraged me to think of a wider audience for my project and posed questions to which I am still looking for answers.

At Temple University, I have enjoyed the support, fellowship, and connection of a community of scholars and students. Susan Wells generously gave her valuable time, sage advice, and expert guidance at every step of this project as it developed and grew to become what it is now. Eli Goldblatt encouraged me at every juncture in this journey and offered his warmhearted counsel to me without reserve. Shannon Miller both inspired and reassured me, helping me to see the path that my early career could take. Joyce Joyce supported me strongly and gave me the example of a leader who listens. Katherine Henry and Sue-Im Lee generously shared their experiences with me and made my way a little easier because of them. Nichole Miller is a trusted friend with whom I am lucky to share stories and laughter. Bill Gonch has been a kind and generous reader of this project, asking astute questions and suggesting valuable strategies for revision. I am deeply indebted to the undergraduate and graduate students with whom I have shared my ideas on rhetorical touch, especially the undergraduate writing students whose experiences are described in chapter 6. I have also treasured connections and collaborations with colleagues at Temple's Institute on Disabilities and the First Year Writing Program.

At Temple, I also enjoyed support in the form of release time and fellowships that were instrumental in this book's completion. Summer Research Awards from the College of Liberal Arts and a Study Leave were invaluable. A Faculty Fellowship provided by the Center for the Humanities at Temple was extremely helpful. I appreciated research assistanship from James Brown and Elizabeth Seltzer. In addition I am grateful for an award provided by Penn State's Rock Ethics Institute during the early stages of this project.

I feel especially appreciative of those close to me who supported me, in ways too many to count, while I worked on this book. I appreciate those who have afforded me the time, space, and patience I needed to finish this project. To my parents, Charles J. Walters Jr. and Elaine M. Walters, my first and most ardent cheerleaders, thank you. To Jeff Fraser, with whom I share my life, my heart, and my home, and whose love, companionship, and support have meant the world, thank you for understanding. I value dearly the love and support of my brother, Casey Walters, my sister-in-law, Kelly Walters, and my nephew, Henry Walters. Heartfelt thanks to Patricia and William Kwasniewski for their love and support. I deeply value the friendship of Liz Kuhn, who always makes me laugh and whose encouragement has been tremendous. I appreciate the support of Jennifer Conroy, who helped me keep perspective at crucial moments during this project, especially during its growing pains. I thank Maria Clark, in many ways my reason for writing this book, for sharing her experience and life with me.

In many ways this book is meant to be a contribution to the work of a growing number of scholars at the intersections of rhetoric and composition and disability studies. I am continually buoyed by the fellowship, liveliness, and generosity of these scholars, many of whose works I admire greatly, including Brenda Jo Brueggemann, Cynthia Lewiecki-Wilson, Jay Dolmage, Margaret Price, Stephanie Kerschbaum, and Amy Vidali. At conferences, on listservs, and on the page, your work has energized me and motivated me to contribute. Thank you. Brenda and Cindy's feedback on this project was particularly helpful and generative; I hope I have done it justice. The book is much the better for the brilliant suggestions and insightful questions they offered.

At the University of South Carolina Press, I am grateful to Jim Denton for expressing interest in this project in its early stages and for the advice and guidance he has provided at all stages of its development and fruition.

Portions of chapter 4 have appeared, in different form, in *Disability Studies Quarterly* and *JAC: A Journal of Rhetoric, Culture, and Politics* and are reprinted with permission and acknowledgment.

Introduction

Rhetorical Touch—Sensations, Bodies, Rhetorics

A simple but generative thought experiment begins *Treatise on the Sensations* by the eighteenth-century philosopher Étienne Bonnot, abbé de Condillac. A marble statue, constructed internally with the exact physical structure of the human body but as of yet insentient, serves as a proxy by which readers are to imagine themselves. Condillac writes, "I forewarn the reader that it is very important to put himself exactly in the place of the statue we are going to observe. He should begin to live when it does, have only a single sense when it has only one, acquire only those ideas that it acquires . . . in short: he must be only what it is" (trans. Philip 155). Sense by sense, starting with smell and ending with touch, Condillac brings the statue to life. His purpose is to "show how all our knowledge and all our faculties come from the senses or, to be more precise, from sensation" (155). He leaves the sense of touch for last because it is the sense on which all of the rest depend, responsible for turning the statue into a body: "Current sensations of hearing, taste, sight and smell are only feelings as long as these senses have not yet been instructed by the sense of touch, because the mind can then take them only for modifications of itself" (167). Touch is "the only sense that judges external objects on its own" (224). In turn it is the only sense capable of "teaching the other senses to refer their sensations to surrounding objects" (213).

To illustrate the importance of touch as a teacher of the other senses, particularly of sight, Condillac describes the experience of a young man born blind, whose cataracts are removed in an operation. Making the point that initially the young man "did not want the operation," Condillac reports that "he did not imagine what he might be lacking. Would I know my own garden better, he asked? Will I walk there more freely? For that matter, do I not have an advantage over others moving about at night with greater confidence?" (290). Condillac dismisses these questions, writing, "In truth he could not long for an advantage that he did not know," but the young man remains stalwart in his reasoning, agreeing eventually to undergo the operation only for a specific reason: to learn how to read and write (290).

Condillac observes that after the operation the young man does not immediately or easily acquire the ability to see. Instead he "learns to see only as a result of study" (292). At first objects appear to him actually to touch the exterior surface of his eye, and he experiences challenges with depth, dimensionality, shape, and size. But eventually touch teaches him to unite his previous experience with his new ability to see: "when he was shown objects that he recognized by touch, he observed them carefully in order to recognize them on another occasion by sight" (292). Although Condillac does not report on the young man's pursuit of literacy after the operation, it is reasonable to assume that touch also aided him in this endeavor, perhaps as much or even more so than sight. Touch is the bridge of experience that fosters a new way of seeing for a previously blind young man, yoking his new experience to his more familiar way of inhabiting his world and affording him the opportunity to read and write in a world in which Braille is not yet available.

For the young blind man, as well as for the readers of Condillac's treatise, touch is transformational, especially in relation to embodiment and language. Sensation is the primary way in which members of Condillac's audience, following his initial directive, read themselves into his treatise. Ideally readers, heeding Condillac's instructions for reading, imagine themselves exactly in the place of the statue, sensing their bodies and their worlds as they read. At the same time the sense of touch ushers the insentient statue into personhood, embodiment, and language, creating an "I" in tactile relation to an external world. Condillac explains that "the statue can say 'I' as soon as some change in its fundamental feeling occurs. Consequently the feeling and the 'I' are in their origin the same thing" (trans. Carr 75). It is "solely by the aid of touch" that the body says something, recognizes itself, and discovers how "the 'I' can be modified in order to discover what it is capable of" (75).

Taking Condillac's suggestion that touch is a valuable resource for discovering what a body may be capable of, I ask in this book what touch, the teacher of the senses, can teach us about rhetoric and the bodies that use rhetoric. In the following pages I explore the ways that touch is rhetorical and argue that understanding touch as rhetorical can reveal new ways of valuing embodiment in relation to rhetoric. Extending Condillac's example of the blind man, I bring the consideration of disability to the front and center of analyses of rhetoric and touch. The blind man in Condillac's treatise gave up the blindness he felt comfortable with in order to gain traditional literacy and the ability to read and write. He did not perceive himself as "lacking" anything in his blindness, but in order to achieve the traditional rhetorics of reading and writing, he underwent surgery during a time when few alternative paths to literacy were available to him. If touch were understood as rhetorical—a method of expression that draws on the sense of touch to communicate, form messages, persuade audiences, convey emotion, establish identifications, and craft character—might the man have felt less pressure to "correct" his blindness? In broader terms, how might understanding touch as rhetorical and rhetoric as tactile change how we think of rhetoric, especially regarding what kinds of bodies and minds have access to rhetorical production and its elements, purposes, and possibilities?

Drawing on the experience and writing of people with disabilities, I argue for an understanding of touch as rhetorical and show that touch is integral to the history, theory, practice, and pedagogy of rhetoric. I define rhetorical touch in this inquiry as a potential for identification among bodies of diverse abilities that takes place in physical, proximal, and/or emotional contact. By foregrounding the rhetorics of disabled people, I theorize touch in relation to the types of bodies that rhetorical history and theory have tended to exclude. By focusing on the potential for identification, I foreground identification as a partial and incomplete process that does not erase differences among people but operates in the potentials of difference. Touch is a rhetorical strategy that can be particularly valuable and meaningful for people with disabilities but that also enlarges the means of persuasion for all rhetors, regardless of ability or disability.[1]

Extending Kenneth Burke's suggestion that rhetorical identification can take place via "common sensations," I explore identification as a facet of rhetorical touch because studying touch reveals how all bodies—disabled or not—are connected and interdependent in physically significant ways. Disability, as both a social and a bodily experience, is integral to rhetorical touch, especially in relation to the possibilities of identification concerning broader conditions of embodiment. As many disability studies scholars have pointed out, everyone, even those who identify as able-bodied, will likely encounter disability in their own lives or in the life of a close relation at least once in a lifetime. The beginning and end of anyone's life will most likely involve dependency, care, and personal assistance. These contingencies are frequently enacted by touch. We have been and likely will be bathed, fed, changed, and clothed, all ministrations that involve close physical contact, interdependency, and trust. Furthermore these contingencies affect everyone, as scholars such as Lennard Davis, Rosemarie Garland-Thomson, and Simi Linton have pointed out, because disability and ability are parts of the same overarching system. Terms or categories of experience such as disabled and nondisabled, normal and abnormal, "normate" and freak are all mutually constitutive, codependent, and unstable.[2] A theorization of the sense of touch contributes to the questions and explorations surrounding common conditions of embodiment and to the destabilization of the categories that purport to separate them.

Exploring rhetorical touch as a potential for fostering partial identifications among people of diverse experiences of embodiment can encourage opportunities for social cohesion, alliance building, and cultural connection among people of different levels of ability and disability. I focus primarily on how people of diverse abilities put their experiences of touch into language, but rhetorical touch is not necessarily uniquely linguistic or always conveyable in written or spoken form. Rhetorical touch takes place when bodies come in contact; the meanings produced by this contact are rhetorical in that they convey messages, craft character, and create emotion in a way that fosters a potential for identification and connection among toucher and touched. In short, touch is rhetorical because it is epistemic, creating knowledge, communication, and understanding about the widest ranges of embodiment and ways of being in the world. Understanding touch as rhetorical makes rhetoric accessible to a wider range of bodies and minds, increasing the means of persuasion and possibilities of rhetoric. Understanding touch as rhetorical redefines rhetoric for the widest range of bodies and minds, and it challenges a tradition in which people with disabilities are often excluded from rhetorical agency.

Defining and Redefining Touch: Embodiment and Disability

The sense of touch is usually understood as a sensory perception that results from a combination of nerve receptors and nerve endings that relay information concerning pressure, temperature, pain, and movement. Touch is also often understood as a sense of communication, as it conveys not only sensory information but also emotional, social, and interpersonal knowledge. The sensory information of touch—the tangible sensation sent along nerves and receptors—and the more affective information of touch—the feelings, emotions, or connections it evokes—are both crucial for understanding a rhetoric of touch. Touch is a sense that involves both the body and the mind and that takes place at the intersection of one or more bodies. Touch, perceived and a perception, is always relational, bringing bodies in contact and creating a new space. Touch, at its most productive, depends on more than one body or a sense of "the body" that exceeds singularity. Even in the case of self-touch, one part of the body connects with another part, forming a third space of contact. This third space is even more pronounced when two distinct bodies connect, as it forms a separate but joined third space for meaning, communication, and experience. Touch is often ambiguous, with no clear line separating the body touched and the body touching. This ambiguous space eschews the easy binaries of subject and object, the disabled body and the nondisabled body, inside and outside, private and public, individual and social, and material and discursive. Accordingly the study of touch is the study of "bodies" rather than the study of "the body."

Frequently explored through the senses of sight and sound, rhetoric is less understood through the sense of touch, even though it is often understood as physical, material, and embodied. Metaphors of touch suffuse common expressions about language and language use, and yet the significance of tactile communication remains underexplored from a rhetorical perspective. We "keep in touch" with someone by speaking or writing to them; we are "touched by" a piece of writing or speech that creates an emotional response; we "scratch the surface" of topics in order to gain new understanding and "delve deeper" into them to explore more. Difficult issues or problems are "thorny," situations are "sticky," and people can be "touchy," "prickly," or "abrasive." As Constance Classen notes in *The Book of Touch,* an abundance of words used as synonyms for "cognition" are tactile, such as "comprehend," "cogitate," "conceive," "grasp," "mull," "ponder," and "ruminate" (5). Etymologically even the verb "to write," from the Greek *grapho,* means "to scratch," a tangible expression of meaning.

Touch ties writing, communication, and expression closely to the bodies from which it arises, and yet the sense remains relatively unexplored.[3] As Nina Jablonski explains in her study of skin, "touch has not garnered the scientific or public attention it deserves, possibly because its influences on human well-being are more subtle than those of the so-called distant senses of sight and hearing" (97). The other four senses correspond to specific areas—eye, ear, nose, and tongue—but the sense of touch, housed in the skin, the body's largest organ, stretches over the entire body. Skin is "the interface through which we touch one another and sense much of our environment" (2). The interface of touch reaches into the body as well, involving stimulation of the skin and resulting in sensations of pressure, vibration, temperature, or pain. As Diane Ackerman describes in her natural history of the senses, "Touch is a sensory system, the influence of which is hard to isolate or eliminate," as it often functions in combination with other senses (77). In addition "touch is difficult to research" in many ways because "scientists can study people who are blind to learn more about vision, and people who are deaf or anosmic to learn more about hearing and smell, but this is virtually impossible to do with touch" (77). A particularly hardy sense, touch cannot be studied by its absence; it is the first sense to develop in the womb and is almost never completely erased by impairment, old age, or debility.[4] When touch is affected by disability, it invites exploration, experimentation, and knowledge making.

Touch is a challenge to study; yet many people with disabilities are writing about touch and documenting its effects on their daily lives. Touch is a particularly productive sense through which to study and value the experience of disability. As Davis points out, "Disability exists in the realm of the senses. The disabled body is embodied through the senses. So there is a kind of reciprocal relationship between the senses and disability. A person may be impaired by the lack of a sense—sight, hearing, taste, or even touch, although touch is almost never completely gone. Yet, paradoxically, it is through the senses that disability is perceived" (*Enforcing Normalcy* 13). This reciprocal relationship of perception and reception between the senses and disability is especially typified by touch, which provides a common ground of experience that almost anybody can mobilize in the pursuit of rhetorical identification. As Mark Paterson notes, touch is a sense that nearly everyone possesses to some degree, regardless of ability or disability: "Unless we have an extremely rare neurological condition, touch is present within every single interaction with objects, and a considerable amount of interaction with people" (2). A wide range of people with disabilities have been actively writing and communicating by drawing on that one sense that is "almost never completely gone" to explore the reception and experience of disability.

Although everyone uses touch, people with disabilities, some of whom may experience impairments in a range of sensory perceptions, often actively employ their senses of touch in diverse ways to relate to others and to establish a sense of their own bodies in the world. Disabled people use touch to connect to other people such as assistance aides or to use assistive technologies, often using touch to develop both autonomous and interdependent relationships. As Janet Price and Margrit Shildrick explain, touch is a "highly significant" form of interaction in the context of disability that produces an array of different results: "What occurs—if there are problems of mobility for example—is that the disabled person may find herself being touched by others in ways that far exceed normative contact, especially between strangers; or if she is deaf or vision-impaired, she equally may need to touch others to gain attention or for orientation and recognition" (70). This relationship to touch among people with disabilities—always complex, never simple—forms what I consider a valuable approach toward the understanding of touch as rhetorical, primed for forming connections with a wide range of disabled and nondisabled people.

Touch is a sense that people with a wide range of physical, psychological, and cognitive disabilities can use to their advantage, forming potential spaces of rhetorical contact and identification with others of varying abilities. Disability often occasions new and diverse approaches to touch. As Petra Kuppers asserts, "Disability is a realm I traverse with a strong sense of the haptic, the touch of concepts and bodies" (225). Kuppers invites the possibilities afforded by a sense of connection based on bodily experience and touch with others. She desires a way of understanding disability in which the "extrinsic and intrinsic mix and merge, as they do in my own physical and psychical being when I am in pain, and cannot walk up the stairs, and wish for a painkiller, and take pride in my difference . . . and feel unable to speak of the nature of my discomfort, cannot find the words, but find comfort in the company of others whose pain may be different, but who somehow feel sympatico" (225–26). This approach to disability "touch[es] words" with "experiences . . . and other concrete objects in the world (stairs, pills, people, the ground, a table around which we are sharing our libations)" (226). This intertwining of experience and bodies is at the heart of rhetorical identification enacted through rhetoric of touch. This sense of sympatico among bodies, places, things, and feelings forms a platform onto which people of different experiences of embodiment can begin to forge connection, identification, and collaboration.

Situating Disability: Touch in a Disability Studies Perspective

Disability is a productive and valuable way of being in the world that can make some people more inclined to use touch rhetorically. To theorize a rhetoric of touch and to realize the fullest potential of touch for the widest range of rhetors, disability and ability must be understood from a disability studies perspective. In taking a disability studies perspective in this inquiry, I follow scholars and activists in disability studies, theory, and activism and define disability not as a personal tragedy, individual flaw, or negative experience that should be pitied, fixed, cured, or hidden away but rather as a valuable and meaningful way of being in the world that is socially, politically, and culturally constructed and constructible. A disability studies perspective supports an understanding of disabilities—cognitive, physical, and psychological—as expressions of human difference and variation that are valuable and significant. Accordingly disability studies theorists and activists adopt the stance that disability is not a "problem" in and of itself but can become an issue primarily because most environments, attitudes, and societies fail to accommodate difference and do not appreciate disability as a source of positive and meaningful experience.[5] "Disability" in this register and throughout this book refers to the wide spectrum of differences, including temporary able-bodiedness, in psychological, cognitive, and physical experience. "Disability" is a term that exists in relation to ability, which I define not as a normal or unchanging bodily state but instead as a temporary, contingent, and diverse bodily state.

As scholars such as Susan Wendell, Eva Kittay, Linton, and Garland-Thomson have explored, the study of disability frequently intersects with feminism and feminist theories of the body. Attention to rhetorical touch extends existing intersections between studies of the body in feminism and rhetoric. As James C. Wilson and Cynthia Lewiecki-Wilson explain, "Since rhetoric has always been categorized as 'embodied' (and thereby valued less than philosophy), it shares with disability studies and feminism a common position and interest in deconstructing this polarity and revaluing 'embodied' theory and scholarship expressive of a standpoint" (7). Eschewing an essentialist or universalist approach to standpoint, disability studies shares with feminism a critically reflexive practice of examining knowledge as specific and socially situated. This self-reflexive perspective in disability studies is often demonstrated in a principle that prioritizes the experiences and perspectives of disabled people, characterized in the mantra of "nothing about us without us."[6] This principle holds that the most responsible way of learning about disability comes from the experiences and contributions of disabled people themselves. In the chapters that follow, I rely on the writings of people with disabilities who are reaching for connections to other disabled and nondisabled people. These efforts toward initiating rhetorical identification, however partial and incomplete, with a wide range of people with diverse embodiment experiences constitute an act that is connective and generative but is also attentive to the many forms of difference that shape experience. Exploring the spaces between bodies in connection, the study of touch deepens areas of common interest with feminism and rhetoric, especially feminist and alternative rhetorics focused on revising a normative and individual tradition, exploring polyvocality and embodiment, and valuing intersectionality with other forms of difference (Lunsford; Glenn; Biesecker; Ritchie and Ronald; Richardson; Royster; Logan).

Touch also extends intersections between disability studies and the study of connections between bodies, discourses, and social practices in other areas. As disability theorists such as Mairian Corker and Tom Shakespeare have shown, the field of disability studies shares common ground with poststructuralist and postmodern social theory. Shakespeare and Nicholas Watson explain, "For us, disability is the quintessential post-modern concept, because it is so complex, so variable, so contingent, so situated. . . . Disability cannot be reduced to a singular identity: it is a multiplicity, a plurality" (19). However, as Davis relates, the interrogations of postmodernism and its aftermath remain incomplete in the context of disability: "The universal subject of postmodernism may be pierced and narrative-resistant but that subject was still whole, independent, unified, self-making, and capable" (*Bending Over Backwards* 26). The study of touch operates in these interstices, extending multiplicity and transitioning from a focus on singular identity to the examination of rhetorical identification between bodies with identities that are physically connected, interdependent, and multiple. Touch functions in the spaces between the tensions of discourse, embodiment, social construction, and materiality and in locations of partial and potential identification that bridge individual experience with social and political connection.

Rhetorical Sensations: Tactility in Empedocles, Aristotle, and Burke

Touch is a sense that transcends bodily boundaries; it demands an approach that also transcends boundaries. The chapters that follow combine Empedoclean, Aristotelian, sophistic, and Burkean rhetorics, establishing potentials for understanding how touch is rhetorical for a wide range of diverse bodies. As Debra Hawhee notes, resurgence of interest in the ancients in the last few decades in rhetoric and composition studies aims to be "connective" rather than "writing history for the sake of history," as scholars have been more interested in finding ways in which the ancients "might help us reframe or reconsider contemporary debates" and "discourse[s] already in circulation" ("Bodily Pedagogies" 142). As John Poulakos has shown in "Toward a Sophistic Definition of Rhetoric," sophistic rhetoric is a rich resource for retheorization. Many scholars have used sophistic rhetoric to inform contemporary issues such as feminism (Jarratt), cultural studies (T. Poulakos;

Welch), postmodernism (Vitanza), and the liberal arts (Atwill), among others. I extend this sophistic inquiry of contemporary issues to the study of disability, specifically by focusing on the teachings of Empedocles, who influenced the sophists, while also exploring the concepts of *mētis* and *kairos.* There is much that separates the theories of the sophists, Empedocles, Aristotle, and Burke, but I find crucial spaces of correspondence in these figures' various treatments of bodies and their suggestions of the tactile properties of rhetoric, which I use to create potentials for rereading the value of bodily difference in rhetoric.

The pre-Socratic fifth-century B.C.E. philosopher, physician, and orator Empedocles of Acragas, called the "inventor of rhetoric" by Aristotle, offers a way of understanding rhetoric as a uniquely tactile entity that circulates among bodies and minds (Diogenes Laertius 8.57–58). Empedocles's theory of pores and effluences posits that all matter is comprised of four elements—air, water, earth, and fire—which are put into physical contact by the opposing but complementary forces of Love and Strife, which he characterizes as "energies." Physical effluences emanate from the four elements, moving through pores of corresponding elements, creating bodies and matter by a process of physical intermingling called the "symmetry of pores." Theophrastus characterizes Empedocles as explaining "all mixture" of living things and inanimate matter as produced by "sensible objects" fitting into one another: "thus everything will be capable of sensation, and mixture, sensation and growth will be the same thing; for [Empedocles] explains everything by symmetry of pores" (quoted in Guthrie *A History of Greek Philosophy* 233).

Empedocles's lesson employing the example of respiration is considered to be representative of his general theory of pores and effluences, demonstrating how sensation is the guiding principle of everything: "All things draw breath and breathe it out again. All have bloodless tubes of flesh extended over the surface of their bodies; and at the mouths of these the outermost surface of the skin is perforated all over with pores closely packed together" (Diels and Kranz 31.B.100; Burnet 219). In keeping with the general theory of pores and effluences, the "bloodless tubes of flesh" that extend over the surface of bodies and matter are perforated with tiny and densely packed "pores" in the skin through which air and other effluences pass. Sensation and effluence are typically understood as the guiding principles of the construction of the cosmos, bodies, and matter in Empedocles's teachings, but lesser explored and understood are the implications of this emphasis on sensation and tactility for rhetoric.

In the chapters that follow, I take Empedocles's suggestion that touch is the "broadest way of persuasion" and examine how touch structures Empedocles's theories of bodies, rhetoric, and pedagogy (Diels and Kranz 31.B.133; Burnet 225).[7] Specifically, I explore Empedocles's concept of *logos,* performed in his own teachings and evinced in Gorgias's rhetoric, arguing for an understanding of *logos* that inherently includes tactile and fleshy properties. Empedocles, in Plato's *Meno,* is reported to have taught Gorgias, his student, the theory of pores and effluences (76C). As Hawhee notes, for Gorgias, "bodies and souls, like bronze and silver, were porous entities that allowed effluents and other substances (words, fire) to pass through" (*Bodily Arts* 79). Gorgias's likening of *logos* to drugs coursing through the body in his *Encomium of Helen* is a description that follows directly from his teacher Empedocles's theories. This understanding of *logos* by Empedocles and Gorgias exceeds metaphor and positions rhetoric in direct tactile relation to bodies. In this view, words physically pierce bodies, moving through pores to distribute various effluences throughout bodies. As with his theory of the symmetry of pores, Empedocles can be understood to view *logos* as operating via proportions that fit into each other in diverse ways to create not only bodies and matter of various shapes and forms but also ways of thinking and communicating that are equally diversely shaped. In this way Empedocles's theories of sensation and rhetoric are similarly based on principles of transmission and effluence, connecting bodies and rhetoric together in flexible ways.

Aristotle, also intensely interested in sensory experience, wrestled with and rejected certain elements of Empedocles's theories but shared with Empedocles an understanding of both sensory experience and rhetoric as energetic, constructive, and transformative. For Aristotle, the study of touch was both perplexing and fascinating. The phenomenologist Jean-Louis Chrétien notes that Aristotle provides "the most radical and patient analysis of touch to be conducted so far in the history of philosophy" (84). This analysis resulted in complex and sometimes contradictory assessments of touch. As Paterson notes, Aristotle's "texts remark variously on the exceptional nature of touch, its primacy, yet also its base nature" (17). Aristotle, calling touch fundamental in *De Anima,* believes that "without touch there can be no other sense" (435a) but questions where the sense of touch resides. He notes that "it is not clear what is the underlying unity of touch, as sound is of hearing" and theorizes that if "one were to stretch a covering or membrane over the skin, a sensation would still arise immediately on making contact; yet it is obvious that the sense-organ was not *in* this membrane" (422b–423a). As Anne Davenport explains, Aristotle anticipates the phenomenological problem of the "veiling" of touch whereby touch gives the impression of immediacy, suppressing the distinction between two bodies or entities. In *De Anima,* in particular, Aristotle meditates on the confusion that touch occasions between bodily boundaries, questioning whether it is one sense or many and what its proper medium can be thought to be.

In the chapters that follow, I take up Aristotle's search for an "underlying unity" of touch and suggest that the concepts of change, transformation, and potential that undergird Aristotle's approach to sensory perception also situate his approach to rhetoric. I argue for an understanding of Aristotle's definition of rhetoric as attentive to people of varying degrees of ability and disability, especially those who use touch as a rhetorical strategy. When Aristotle states, "Let rhetoric be [defined as] an ability, in each [particular] case, to see the available means of persuasion" (I.2.1355b1, in *On Rhetoric,* trans. Kennedy, 36), it is possible to understand "ability" not as exclusive to able-bodied rhetors but in a dynamic and energetic sense, inherently inclusive of rhetors with a wide range of abilities and disabilities. Drawing on Aristotle's definitions of art in his *Nicomachean Ethics* and his classification of the senses in *De Anima,* I explore similar properties of transformation and change in each, connecting these definitions to the potentials of rhetoric, particularly to the artistic proofs of *ethos, pathos,* and *logos.* In doing so, I position rhetorical touch as a potential based on the wide range of abilities and disabilities that different rhetors bring to rhetorical production and rhetorical identification. Extending Aristotle's classification of rhetoric as a *dynamis,* or a potential or capacity, I situate touch as an available means of persuasion for a wide range of rhetors.

Like Empedocles and Aristotle, Burke uses sensation to forward a key concept in his rhetoric. Specifically, Burke invokes sensation to define his concept of rhetorical identification, mobilizing a transformative understanding of rhetoric to explore how people of different interests and experiences can come together. Burke describes identification, writing, "You persuade a man only insofar as you can talk his language by speech, gesture, tonality, order, image, attitude, idea, *identifying* your ways with his" (*A Rhetoric of Motives* 55). Language, in this definition, stretches broadly to encompass a "way of life" and an "*acting-together*" that includes elements ranging from speech, tone, and gesture to "common sensations, concepts, image, ideas, attitudes" that make people "*consubstantial*" with each other (21). As I explore in successive chapters, "common sensations" such as touch are particularly productive means of consubstantiality and identification.

In the afterword to *Attitudes toward History,* Burke describes the interaction between sensation and language as a kind of duplication:

> I refer to a kind of duplication that arose when our primeval ancestors, by learning language, no longer experienced a sensation solely as a sensation. For instance, when they touched something that *felt hot,* their newfound ways with language enabled them to *duplicate* the *sensory* experience in the "transcendent" terms of a *nonsensory* medium such that our aforesaid primeval ancestors could say *"That feels hot."*
>
> And precisely at that time here on Earth the realm of *Story* entered the world. The taste of an orange *is a sensation.* The words "the taste of an orange" *tell a story.* And the story they tell is such that it must be a somewhat different story, depending on whether the hearer has or has not tasted an orange. Once such

> words have arisen as terms for *sensations*, their use can be extended *attitudinally* to encompass such steps as the one from "That feels warm" to "He, or she, is warm-hearted." (382–83)

Burke's example of our primeval ancestors' transition from sensation to language anticipates a more recent argument by the psychologist Robin Dunbar, who posits that the evolution of human language depends on touch. According to Dunbar, human language evolved from the practices of our ancestral apes, who used physical contact such as grooming to relay social information, initiate bonding, and maintain connection among group members. In his view, language evolved to replace the social rewards of physical grooming through touch once groups became too large and grooming became overly time-consuming (190).

For Burke, the "realm of *Story*" charts this evolution from sensation to language. As Burke describes, words tell a story that may be related to sensations, but these same words may also tell a different story, depending on who has touched, tasted, or seen what. To say that something feels warm is different from saying that someone is warmhearted, although these statements are attitudinally connected through sensation and symbolism.

Similar to how words tell a story, sensation also tells a story. As much as words may duplicate sensations, they also may leave out information that sensation conveys. Different interests and attitudes are at stake in these different stories. As Burke relates, "Storytelling words designed to duplicate the wordless aspects of our environment greatly expand the range of *attitudes* by which we relate to one another in keeping with the clutter of concordant and discordant interests *socially rife among us*" (*Attitudes toward History* 384–85). A range of different attitudes comprising various levels of accord and discord directs the stories that we tell about our environments and is subject to the varying interests at stake in a particular social situation, forming identifications and divisions.

In this register, words and their "wordless" counterparts—sensations—are intimately connected to the basic elements of the Burkean concept of identification. Burke outlines identification as the "intermediate area of expression that is not wholly deliberate, yet not wholly unconscious," which "lies midway between aimless utterance and speech directly purposive" (*A Rhetoric of Motives* xiii). Sensation and the stories we tell about sensation also often reside in this "intermediate" realm of expression between utterance and speech and are subject to the various kinds of attitudes that connect and divide, as well as to specific contexts, cultures, and social pressures. A rhetorical study of touch contributes to the understanding of these attitudes and expressions. In particular, touch and the stories it tells have the potential to connect and divide people among various "concordant" identifications and "discordant" disidentifications.

Burke evokes both the figure of Empedocles and the rhetoric of Aristotle in his construction of identification as a key concept for extending the "range of rhetoric." With identification, Burke seeks to develop a kind of rhetoric that does not just attempt to achieve traditional or classical persuasion—"the holds and the counter-holds, the blows and the ways of blocking them, for every means of persuasion" (*A Rhetoric of Motives* 52)—but also exists to connect people in new and diverse ways. Through an analysis of Matthew Arnold's poem "Empedocles on Etna," which tells the legend of Empedocles's dramatic throwing of himself into a volcano, Burke reminds us that the "imagery of killing is but one of the many terminologies by which writers can represent the process of change" (xiii). Identification is an "instrument" that can initiate this transformation, a way of "showing how a rhetorical motive is often present where it is not usually recognized, or thought to belong" and a method for "rediscover[ing] rhetorical elements that [have] become obscured" (xiii). Touch functions similarly, operating as a vehicle for identification and transformation, bringing diverse bodies together.

As Burke suggests, the stories that sensation tells about rhetoric and rhetorical identification are relevant to expanding the range of rhetoric and for initiating transformation. Similar to how sensation tells a story, touch too tells a story about the possibilities and potentials of rhetoric. In particular, tracking touch in the rhetorical tradition uncovers a story of a tangled relationship between touch, bodies, disability, and rhetoric. Exploring this relationship reveals the limiting ways in which the tradition has shaped certain bodies for rhetoric but also the more expansive possibilities for valuing the widest range of bodies and minds capable of initiating rhetorical identification and transformation.

Shaping the Able Body: Tangles of Touch in the Tradition

The story of touch as a divider, a force of division and harm, is well known. Touch is a sense that creates special anxiety in the rhetorical tradition. The ancient goddess of persuasion, Peitho, was also the goddess of seduction and charming speech, and when combined with force (*bia*), she represented rape; her attributes included a ball of binding twine. She is often pictured with her hand lifted in persuasion or fleeing the scene of a rape, and her presence conjures the danger and threat of touch, especially when combined with persuasion.[8] As Gorgias explains in his *Encomium of Helen* with a vivid explanation of how Helen may have been constrained by words, powerful persuasion can operate as physical force that can trap, entangle, and constrain. This potential power of rhetoric is most keenly felt in its effects on the bodies, minds, senses, and emotions of audiences. The senses of sight and touch are especially involved in this power and potential of rhetoric. In Gorgias's estimation, "sight engraves upon the mind images which have been seen. And many frightening impressions linger, and what lingers is exactly analogous to [what is] spoken"

(11.17; trans. Kennedy 54). In keeping with his teacher Empedocles's theory of pores and effluences, Gorgias's *logos* possesses uniquely tactile properties, "engrav[ing]" images and "impressions" of fear or other feelings onto listeners' minds and epitomizing the power, threat, and potential of rhetoric.

This tangled relationship between touch, rhetoric, and disability in the rhetorical tradition is in large part a result of complex and often contradictory and longstanding connections between the senses, embodiment, and rhetoric. In "Studying Disability Rhetorically," Brenda Jo Brueggemann and James Fredal note that "throughout its 2500 year history, rhetoric has never been particularly friendly to the disabled, the deformed, the deaf or mute, the less-than-perfect in voice, expression or stance" (251). Rhetorical training and delivery demand a well-functioning body and mind; "energy, willful self-control, and physical, intellectual and financial resourcefulness" are ideal, and "anyone who hoped to control the will of an audience had first to control their [*sic*] own voice and body" (251–52). Implicit in this control over the body and its senses is the assumption of ability and independence.

Contradictions abound in this assumption of ability regarding the role of the body and the senses. As Brueggemann and Fredal point out, "Though rhetoric required perfectly functioning bodies, its detractors also condemned it for appealing to the body and the senses at all" (252). As a result any "bodily deformity thus at once prevented any rhetorical achievement while, at the same time, it symbolized the problem with rhetoric as a deceptive and sensuous art" (253). In this view, rhetorical achievement demands only certain kinds of appeals to and uses of the senses by certain bodies—normal, able bodies that use the senses in nonthreatening and prescribed ways.

In response to this directive, the rhetorical tradition shapes a particular kind of rhetor—one who is usually singular, able-bodied, independent, and normal. Disability is often the raw material against which the ideal rhetor is shaped. This process of shaping and training the rhetor often takes on physical, material, and even tactile properties. The Roman rhetoricians Cicero and Quintilian, often seen as key synthesizers of the ancient tradition, both require an able orator, in the form of a body and mind shaped by discipline, training, and practice. In *De Oratore,* the "gifts of nature" and "inborn capacity" are chief in Cicero's requirements for the ideal orator; among these that Crassus details are "the ready tongue, the ringing tones, strong lungs, vigor, suitable build and shape of the face and body as a whole" (1.25.113–14; trans. Sutton and Rackham 81). This ideal orator forms a pattern for the rhetorical training of bodies based on expectations of ability and normality.

The shape of this able body of the rhetor starkly contrasts to that of disabled people, who are disqualified as rhetors in Cicero's estimation. "There are some men either so tongue-tied, or so discordant in tone, or so wild and boorish in feature and gesture, that, even though sound in talent and art, they yet cannot enter the ranks of orators" (1.25.115; trans. Sutton and Rackham 81). Cicero's disqualification echoes Plato's parable of the two horses in the *Phaedrus,* which separates able and disabled bodies through a process of physical and tactile training and shaping. One horse is "upright and has clean limbs; he carries his neck high, has an aquiline nose, is white in color, and has dark eyes . . . he needs no whip, but is guided only by the word of command and by reason" (253D–E; trans. Fowler 495). The other horse, however, "is crooked, heavy, ill put together, his neck is short and thick, his nose flat, his color dark, his eyes grey and bloodshot; he . . . is shaggy-eared and deaf, hardly obedient to whip and spurs" (253E; trans. Fowler 495). The "crooked" horse is eventually tamed, trained, and shaped by the charioteer's repeated pulling on the bit, which causes pain and "covers his scurrilous tongue and jaws with blood" (254E; trans. Fowler 497). This bodily discipline translates into rhetorical discipline. For Socrates, "every discourse must be organized, like a living being, with a body of its own, as it were, so as not to be headless or footless, but to have a middle and members, composed in fitting relation to each other and to the whole" (264C; trans. Fowler 529). The parable of the two horses, especially the respective form and shape of the upright horse, suggests a rhetoric and a rhetor who are similarly "fitting" and able.

This shaping of the able rhetor also forms a pattern for the bodies and minds involved in rhetorical training and education. Cicero, for example, identifies the ancient rhetor Demosthenes as the ideal orator because he "surmounted natural drawbacks by diligent perseverance: and though at first stuttering so badly as to be unable to pronounce the initial R. of the name of the art of his devotion, by practice he made himself accounted as distinct a speaker as anyone" (1.61.260–61; trans. Sutton and Rackham 191–93). Perhaps as the original "overcoming disability" story, Demosthenes famously retrained himself to speak without a stutter by literally reshaping his body. Cicero explains that "it was his habit to slip pebbles into his mouth, and then declaim a number of verses at the top of his voice and without drawing breath, and this not only as he stood still, but while walking about, or going up a steep slope" (1.61.261; trans. Sutton and Rackham 193). Demosthenes's example functions as a pattern for all potential rhetors in Cicero's "school course in rhetoric," a training by the formation of a "habitual method" (1.31.137; trans. Sutton and Rackham 97; 1.30.135; trans. Sutton and Rackham 95). This method, for which the rehabilitated Demosthenes is the exemplar, involves "the control and training of voice, breathing, gestures and the tongue itself," which "call for exertion rather than art; and in these matters we must carefully consider whom we are to take as patterns, whom we should wish to be like" (1.34.156; trans. Sutton and Rackham 107). Patterning requires a process of shaping based on ability that excludes disability and difference.

Quintilian reinforces this shaping of the able rhetor, especially by comparing rhetorical training to gymnastics and the art of sculpture. He advises the teacher of rhetoric to shape and train bodies and minds in the spirit of athletic training: "As a master of palaestric exercises, when he enters a gymnasium full of boys, is able, after trying their strength and comprehension in every possible way, to decide for what kind of exercise ought to be trained; so a teacher of eloquence, they say, when he has clearly observed which boy's genius delights most in a concise and polished manner of speaking, and which in a spirited, or grave, or smooth, or rough, or brilliant, or elegant one, will so accommodate his instructions to each, that he will be advanced in that department in which he shows most ability" (2.8.3–4; trans. Watson 122–23). The adjectives Quintilian uses to describe the various abilities of the student—smooth, rough, brilliant, and grave—employ tactile properties to indicate aptitude and potential.

He elaborates by categorizing students' bodies and minds through more explicit comparisons between rhetorical training and the art of sculpture. "We must so far accommodate ourselves, however, to feeble intellects, that they may be trained only to that to which nature invites them; for thus they will do with more success the only thing which they can do. But if richer material fall into our hands, from which we justly conceive hopes of a true orator, no rhetorical excellence must be left unstudied" (2.8.12; trans. Watson 124). Students of various abilities are analogous to the raw materials of sculpture for Quintilian—good students are "richer materials" for shaping by the hands of the teacher, for whom no effort should be spared, while "feeble" students are to be given only minimal attention. Quintilian draws out this analogy, sharpening his emphasis on the formation and shaping of the orator: "Had Praxiteles attempted to hew a statue out of a millstone, I should have preferred to it an unhewn block of Parian marble, but if that statuary had fashioned the marble, more value would have accrued to it from his workmanship than was in the marble itself" (2.19.3; trans. Watson 161). Quintilian emphasizes the value of strong natural material, which is only made stronger by efforts to shape it. He relates this relationship between material and work in his characterization of speech as "not the material, but the work; as the statue is the work of a statuary; for speeches, like statues, are produced by art" (2.21.1; trans. Watson 165).[9] This art, however, as Quintilian's previous comments indicate, depends significantly on the strength of original material with which one starts and the level of ability a teacher judges a student of oratory as possessing.

Quintilian's system of rhetorical education echoes Isocrates's pedagogy, which also outlines a shaping process of the body and the mind. As Hawhee has argued, Isocrates and other sophists' rhetorical pedagogies consisted of "bodily arts," which molded and shaped bodies in certain ways. For Isocrates, rhetoric and athletics are parallel arts: "For when they take on pupils, the physical trainers instruct their followers in the postures (*ta schêmata*) that have been invented for bodily contests, while those whose concern is philosophy pass on to their pupils all the structures that discourse (*logos*) employs" (Hawhee "Bodily Pedagogies" 151). *Schêmata* is a wide-ranging term of both bodily and discursive applications, referring to a wrestling move, a figure of speech, a style or manner, or gesticulation (Liddell and Scott). For Isocrates, repeatedly practicing these *schêmata,* or postures, of thinking and being is necessary for students "in order that they may grasp them more firmly and bring their theories in closer touch with the occasions for applying them" (*Antidosis* 184; trans. Norlin 291). Isocrates uses a tactile vocabulary of grasping, firmness, touch, and closeness to describe the desired result of rhetorical training, which is to shape the minds of students in the same way that "bodily contests" of strength and ability shape bodies. This ancient program for rhetorical training is in keeping with similar strategies—from Plato to Cicero and Quintilian—focused on shaping, sculpting, and molding the rhetor to achieve the height of his abilities, eliminating disability to produce the ideal communicator.

Rhetorics beyond the Able Body

A tactile sense of shaping and training has carved out a rhetor of able body and mind in the tradition, but it is also possible to track a different approach to touch in the tradition and in current use. Empedocles, Aristotle, and Burke suggest tactile rhetorics of sensation based on the potentials of generative, transformational, and capacious approaches to touch that are flexible for a wide range of rhetors. In addition people with disabilities writing and communicating about disability who use touch as a rhetorical strategy resist a rhetorical tradition in which the singular, able, and independent rhetor is valued and preferred. The disability activist and performer Cheryl Marie Wade, for example, embodies a different ideal. Wade describes herself, declaring, "I'm a sock in the eye with a gnarled fist" ("I Am Not One of The" 411). Similarly in her poem and performance "My Hands," she uses tactility to shape an alternative kind of speaker:

> Mine are the hands of your bad dreams.
> Booga booga from behind the black curtain.
> Claw hands.
> The ivory girl's hands after a decade of roughing it.
> Crinkled, puckered, sweaty, scarred,
> a young woman's dwarf knobby hands
> that ache for moonlight—that tremble, that struggle
> Hands that make your eyes tear.

My hands. My hands. My hands
that could grace your brow, your thigh
My hands! Yeah! (*Vital Signs*)

Wade moves from visual imagery to tactile actions in her poem, ending with a celebration of her hands that initiates a physical connection to her audience—a touch on the brow or more intimately on the thigh. In a live performance of her poem in *Vital Signs,* she suggestively caresses her thigh when speaking these lines. This rhetorical touch communicates celebration of her differently shaped hands and suggests the possibility of rhetorical identification and disidentification it may entangle. Rhetorically she is both agent and audience, sender and receiver of a tactile message, mixing up categories of speaker/listener, viewed/viewer, and toucher/touched in her physical demonstration of touch. Her repetition of ownership—"My hands"—is sandwiched between a touch that is suggestive and agential. Her final word—an emphatic "Yeah!"—communicates acceptance of her own message, inviting audience agreement. In contrast to the normal, able, and independent rhetor preferred in the tradition, Wade establishes physical connection with her audience via the touch and shape of her nonnormative body.

This transgression of boundaries, both bodily and rhetorical, is akin to what Sharon Crowley imagines when she questions the limits of bodies and the limits of rhetoric. Wondering about the body's edges, she muses, "Where, for example, does the 'outside' of the human eye end or begin? At the eyeball? The iris? The retina? But then where does the retina begin and end?" (*Rhetorical Bodies* 360). Moving toward a formulation beyond "the body," she wonders, "when I place two parts of my body together, say, touching thumb to index finger, both digits experience the touch as both 'inside' and 'outside.' To distinguish skin as the differentiating organ requires intense concentration, as well as my (learned) assumption that the 'skin' of my thumb is the same 'skin' that covers my finger. Is the mouth an 'inside' or an 'outside'?" This series of questions that Crowley asks prods understanding of the body away from singularity and toward multiplicity. The body is less a singular entity and more accurately described as inhabiting various points of contact among itself, other bodies, entities, and the world. This understanding invites alternative construction and shapings of the rhetorical body. Reformulations such as these move rhetoric from the limits of the singular body to the possibilities afforded by bodies in contact. Playing with the edges of the bodies of rhetoric, both Wade and Crowley gesture toward the alternatives that the sense of touch offers for reimagining the bodies of rhetoric.

My own investigation into the rhetorical potential of touch originated in similar questions regarding alternatives to the bounds and bodies of rhetoric that I encountered when collaborating with a disabled woman in the writing of her life story. Shortly after September 11, 2001, I, like many other people, found myself wanting to participate in meaningful work and connect to others in a productive way. I answered a want ad in a local newspaper placed by a woman with a disability looking for a volunteer to aid in the writing of her autobiography. When I met Maria, she explained that she was born with a condition called *osteogenesis imperfecta,* or brittle bone disease. At three feet tall and with particularly bowed limbs, she had difficulty with some everyday tasks but was mainly able to live independently. She tired easily from writing, however, whether with pen and paper or by sitting at a computer, and needed someone to help her with the physical process of writing. She also expressed interest in brainstorming and planning collaboratively.

As Maria and I collaborated, I learned about her life and her work. A frequent public speaker on the subject of disability and employment, she often emphasized the similarities she shared with nondisabled people, but always with an attention to difference. Whether it was in a public speaking role, in her capacity as an employee, or in her everyday interactions with people in public life, her main purpose was to connect with others. For example, when she encountered a curious child at work or in her daily travels, she sought to connect and educate, often using embodied rhetorical strategies. She looked at encounters such as these as chances to explain disability and to educate people, especially children, on the diversity of human forms they would likely encounter throughout their lives. She once held up her hand to a young nondisabled boy who approached us, encouraging him to hold his hand up to meet hers, showing him that they were both alike and different. Rhetorically, Maria was trying to achieve a kind of partial identification with the boy, one that rested somewhat paradoxically on similarity and difference. She displayed her hand and invited the boy to touch her own to let him know that there could be connection amid this sameness and difference.

Touch and the stories told about touch are never neutral. The story I tell about touch and embodiment in this book is situated in a particular experience and perspective, particularly the experience I, as a temporarily able-bodied woman, had writing with a disabled woman. While working with Maria, physically assisting her with the embodied process of writing and brainstorming with her over writing strategies, I experienced a partnership based on a mixture of difference and identification, interdependence and autonomy, trust and experimentation, and individual effort and shared collaboration. Touch calls into question the accepted model of the able, autonomous, independent rhetor in the tradition and disturbs the boundaries and bodies of rhetoric. When I worked with Maria, writing with her and sometimes for her, our physical connection in particular chafed against a tradition in which the prized and preferred rhetor is nondisabled, independent, and single-bodied. When Maria attempted to identify with the curious child, she used her body to communicate a message that resisted assumptions about what and who typically constitute a "normal" rhetor. Understanding rhetorical bodies as those shaped in contact with each other in diverse and interdependent ways resists a rhetorical tradition that desires to mold a normal and able rhetor.

Extending Intersections

To explore touch as a rhetorical art, I both depend on and extend formative studies at the intersections between rhetoric and composition and disability studies by scholars such as Brueggemann, Wilson and Lewiecki-Wilson, Patricia Dunn, Robert McRuer, Jay Dolmage, and Margaret Price, among others. In particular I respond to suggestions by theorists who seek alternative rhetorics not only for people with disabilities but also for the wide range of bodies and embodiments that produce rhetoric. As Lewiecki-Wilson writes, "We need an expanded understanding of rhetoricity as potential, and a broadened concept of rhetoric" that includes a wide range of differently speaking and nonspeaking rhetors ("Rethinking Rhetoric" 157). Toward this end Dolmage imagines that in "recover[ing] a different rhetorical body" for ancient rhetoric, we can value "bodily difference as generative of meaning" ("Breathe Upon Us an Even Flame" 119, 122). Additional and alternative models of the "good man speaking well," as Brueggemann argues in *Lend Me Your Ear,* are useful for valuing the wide ranges of ways in which rhetors speak, sign, and otherwise use "available means" to convey messages. In *Rhetorical Touch,* I position touch as an available means of persuasion, a potential rhetoric for valuing a wider range of bodies, and for exploring the range of rhetorical bodies that make and share meanings.

Understanding touch as rhetorical extends recent studies at the intersections of disability and rhetoric and contributes to the area of body studies in general in rhetoric and composition. Recent studies of the senses in rhetoric and composition have tended to be dominated by attention to sight and sound.[10] Exploring touch, which frequently is experienced in conjunction with sight and sound, more comprehensively extends efforts toward understanding rhetoric as a fully embodied and sensory art. Examining how people with disabilities use touch rhetorically also augments already existing explorations of D/deaf and blind rhetorics as well as rhetorics of people with physical disabilities, which often focus on how disability is visually registered.[11] By exploring touch as a rhetoric used by a wide range of people with both "visible" and "invisible" disabilities—cognitive, psychological, and physical—to connect with a wide range of disabled and nondisabled audiences, I position touch as a particularly productive way to understand diverse bodies in relation and to shift from visuality and aurality to other forms of engagement.

An examination of rhetorical touch contributes to ongoing projects investigating embodiment in rhetoric and composition. Although numerous studies of "the body" in rhetoric and composition have demonstrated the importance of embodiment and physicality in communication, studies of bodies, especially bodies coming together in physical contact, are rarer.[12] In response to theorists such as Gloria Anzaldúa and Mary Louise Pratt, generative attention has been paid to various borderlands and contact zones of writing in face-to-face, electronic, and mediated environments, and yet less attention has focused on what is perhaps the most basic intersection of communication: the physical and material spaces of touch. Collaborative composition models that situate the activity of writing as a socially constructed act shared among various communicators and audiences abound in rhetorical studies; yet few take a materialist turn as specific as touch. As Lisa Ede and Andrea Lunsford, exploring collaboration and authorship, write, despite numerous interrogations, "the question of what it means to acknowledge the death of the author and to proclaim the advent of distributed selves is hardly resolved" (354–55). Accordingly they seek to take a materialist approach to critiquing the "ideologies of the academy [that] take the autonomy of the individual—and of the author—for granted" (357). Questions regarding the autonomy of the individual and of the author are particularly relevant in the context of disability and to the practices of communicating about disability, especially when touch operates as a rhetorical strategy.[13] Touch further disrupts the expectation of a singular, autonomous, and individual rhetor and furthers the model of distributed selves in communication.

Disability studies theorists and activists prize interdependence over autonomy, preferring a model of disability that is social rather than individual. Accordingly disability rhetorics based on touch are produced among bodies in physical and emotional contact and nurtured in the spaces in between autonomy and interdependence. The experience of disability is a social one; as Kuppers, Wade, and Price and Shildrick attest, this experience is replete with physical contact, connection, and interdependence. As I experienced in my work with Maria, a partnership in communication based on the physicality and interdependency of touch does not fit easily into existing models of collaboration, authorship, or agency. If her autobiography were published, simply to state that we were "cowriters" or "collaborators" would not capture the complexity of our experience. The chapters that follow plumb the depths of that complexity, finding a way to value and explore the tactile rhetorics of a wide variety of people of different abilities and disabilities.

Touch, while often subtle, is a means of communication that can be studied and that has a history, theory, and practice in rhetoric. Although the able, autonomous, and independent body has been shaped and preferred in the rhetorical tradition, the study of touch reveals a means of persuasion and identification available to a wider range of rhetors of diverse abilities. Rhetorical touch is based on the capacity and potential of rhetoric and of bodies, with connection and identification, however partial, operating as its purpose. I position touch as a rhetorical art that operates at the intersections of bodies, mobilizing tactile elements of Aristotelian, Empedoclean, sophistic, and Burkean rhetorics for the redefinition of rhetoric for the widest range of bodies and minds.

In chapter 1 I argue for a richer rendering of Aristotle's definition of rhetoric and his triad of appeals by exploring his theories on sense perception and art. Toward that end, a theory of a rhetoric of touch transitions the simplified rhetorical triangle or rhetorical situation (often a limited shorthand of Aristotle's *pisteis*) to an understanding of rhetorical identifications and the dynamic potentials of touch. Drawing on Burke's suggestion of common sensations as a way to induce identification among bodies, I reread his concept of identification through the lens of Empedocles, whom he invokes to describe how bodies function in identification. Using the case of Helen Keller's relationship to touch as an example that extends Aristotle's categorization of rhetoric as a *dynamis*—a potential based on bodies in contact—what emerges is an alternative model for understanding touch in relation to multiple bodies.

I extend this study of identification to chapter 2, arguing for an approach to identification in rhetorical studies and disability studies based on touch that is attentive to material and discursive as well as individual and social registers of embodiment. I examine demonstrations for Section 504, legislation often called the birth of the disability rights movement because it extended civil rights for people with disabilities, in relation to the problem of identification among people of different disabilities, nondisabled people, and temporarily able-bodied people. By exploring theories on touch by phenomenologists and deconstructionists such as Maurice Merleau-Ponty, Jean-Luc Nancy, and Gilles Deleuze, I locate the consubstantiality of touch in rhetorical identification as an act that brings bodies together and reread problems and legacies of identification stemming from 504 through this lens. More broadly, I use conceptualizations of touch as an act of intertwining, delimiting, and folding to move from questions of identity to the possibilities of rhetorical identification and to catalyze revisions of the rhetorical appeals.

In the next three chapters, I take up each of the appeals in relation to different issues of identification in specific contexts of ability and disability, including psychological, cognitive, and physical conditions. In successive chapters I ask what are the *logos, ethos,* and *pathos* of touch? In chapter 3 I explore the limits of *logos* for rhetors with psychological disabilities and those with whom they attempt to identify. I redefine approaches to *logos* narrowly focused on rationality and reason through Empedocles's teaching of a felt *logos* of proportion and show how he used the practice of repetition in his teachings to mold his listeners' minds as flexibly as he molded their bodies. I then reread this felt *logos* through the writings of people with schizophrenia and depression, showing how in online support forums they use repetitive acts of writing based on touch to shape spaces for identification in their real and online lives.

In chapter 4 I identify the limits of *ethos* in traditional and contemporary rhetorical theories for nonneurotypical rhetors who produce rhetoric outside of the familiar confines of the singular body. Specifically, I focus on rhetors with autism, such as Sue Rubin, Tito Mukhopadhyay, Dawn Prince-Hughes, and Temple Grandin, many of whom depend on a complicated relationship to touch in order to communicate or who use facilitated communication or other mediated types of rhetoric. Redefining the sophistic rhetorical practice of *mētis*—cunning intelligence—as a tactile rhetoric based on habit, I explore connections between *mētis* and *ethos,* positioning *mētis* as a way to revise *ethos* to include autistic rhetors who use touch as a rhetorical strategy. For the autistic rhetors I examine, *mētis* is a tactile practice of habituating a sense of *ethos* in relation to other people.

In chapter 5 I grapple with the limits of appeals to *pathos* that frequently accompany stereotypes of the disability experience, especially emotional appeals related to pity, hope, and inspiration, in narratives about physical disability. Exploring writing from disabled rhetors such as Harriet McBryde Johnson, Eli Clare, Kenny Fries, Nancy Mairs, and John Hockenberry, I reread the sophistic practice of *kairos*—opportune timing—as a tactile rhetoric based on the proximity of bodies in contact and relation. This tactile, proximal *kairos,* I argue, presents opportunities for identification and interaction among people of a range of abilities beyond the limited emotions of pity, hope, and inspiration.

Taken together, these three chapters explore actions that bring minds, bodies, and affects together in contact through repetition, habit, and proximity, revising the appeals as flexible for a range of rhetors who use the tactile practices of *logos, mētis,* and *kairos* to achieve connection. More broadly, each of these chapters forwards a *technē* of touch by positioning the critical practice of touch as rhetorical, haptic art.

Chapter 6 tackles the challenges and rewards of teaching disability and technology in the writing classroom. I report on the results of an Institutional Review Board-reviewed classroom study that examines the history of haptics and modern touch technology in relation to the study of ability and disability. Specifically, I describe how I led students to explore the limits of their definitions of accessibility, disability, and ability in the contexts of discourses and practices of haptic technology. Students learn to be more critical of the haptic technologies with which they interact every day, such as the iPad and the iPhone, and to think about how touch is often an unexplored, untheorized, and uncritically accepted sense in discourses of technology and in everyday life.

I conclude by asking what kinds of preconceptions we must let go of and hold on to in order to forward the study of touch in rhetoric, especially in relation to ethical questions. Ultimately touch invites ways of thinking and feeling beyond traditional boundaries. In addition to thinking and feeling beyond the binaries of ability and disability, touch models different modes of engagement beyond other traditional boundaries, including those related to species, language use, and emotion. At its most useful potential as a rhetoric, touch calls into question and disturbs definitions and preconditions for rhetoric itself.

1

Defining a Rhetoric of Touch

Bodies in Identification

In what has become one of the most poignant scenes of disability and literacy in the cultural imagination, Helen Keller describes how the sense of touch initiated her into the world of language. While Keller was walking with her teacher, Anne Sullivan, to a well nearby her house, Sullivan thrust her hand into the water. "As the cool stream gushed over one hand, she spelled into the other the word water, first slowly, then rapidly. I stood still, my whole attention fixed upon the motions of her fingers. Suddenly I felt a misty consciousness as of something forgotten—a thrill of returning thought; and somehow the mystery of language was revealed to me. I knew then that 'w-a-t-e-r' meant the wonderful cool something that was flowing over my hand. That living word awakened my soul, gave it light, hope, joy, set it free! There were barriers, still, but barriers that could in time be swept away" (Keller *The Story of My Life* 35).

In this communication breakthrough, Keller, at the age of seven, finally connected to her world through language. Previously she had learned only to mimic the finger motions of her teacher in what she called a "monkey-like imitation" and understood little connection between objects and their names (*The Story of My Life* 35). After feeling the water, however, Keller understood the connection between "living" words and their names. Although Keller portrays the incident as a miraculous breakthrough, she had previously used a loose method of gestural language and finger spelling. As Georgina Kleege explains, "The pump moment was less a miraculous revelation than a shifting of gears, allowing [Keller] to accelerate, but on the same path [she had] already been traveling" (7). In other words, Keller's instinctual stabs at the connection between language, touch, and her world became cemented in her mind after Sullivan's use of the well water.

After the experience at the pump, touch became solidified as the primary way in which Keller learned the world. She writes, "I did nothing but explore with my hands and learn the name of every object that I touched; and the more I handled things and learned their names and uses, the more joyous and confident grew my sense of kinship with the rest of the world" (*The Story of My Life* 37). Keller's language acquisition progressed to a visceral form of handwriting—finger spelling transmitted hand to hand—as she and Sullivan spelled into each other's hands for hours daily. This learning style shaped Keller's way of thinking and being for the rest of her life. Keller reflects, "In all my experiences and thoughts I am conscious of a hand. Whatever moves me, whatever thrills me, is as a hand that touches me in the dark" (*The World I Live In* 10). Decades after Sullivan's death, she wrote, "To this day I cannot 'command the uses of my soul' or stir my mind to action without the memory of the quasi-electric touch of Teacher's fingers upon my palm" (Keller *Teacher* 51–52).

Touch became Keller's way of writing about the world, but not without ambivalence. Keller exhibited discomfort and even shame regarding her dependency on touch, especially when she learned new ways to communicate, first through speech and then via specific writing technologies. Writing about herself in the third person, she describes the guilty pleasure of finger spelling: "Helen sinned . . . by spelling constantly to herself with her fingers, even after she had learned to speak with her mouth" (*Teacher* 50).[1] Vacillating between the first person and the third person, she continues, "I determined to stop spelling to myself before it became a habit I could not break, and so I asked her to tie my fingers up in paper. . . . For many hours, day and night, I ached to form the words that kept me in touch with others, but the experiment succeeded except that even now, in moments of excitement or when I wake from sleep, I occasionally catch myself spelling with my fingers" (50). Even after she learned to use a typewriter and other information technologies, Keller continued to feel the energy she gained from finger spelling with Sullivan, experiencing the ambivalent stimuli of pain and joy from the memory of communicating exclusively via hand-to-hand touch.

For Keller, touch blurred the boundaries between self and world, often in problematic ways. As Jim Swan describes, Keller's "lifelong experience of *touching* the world in order to 'see' it makes it difficult for her to sustain a sense of boundaries routinely respected by people with hearing and sight" ("Touching Words" 323). Keller especially struggled to understand where her body ended and her teacher's began.[2] In recounting the story of her life with Sullivan, she states, "I feel that her being is inseparable from my own" (*The Story of My Life* 47).

This inseparability surfaces especially in Keller's writing, posing challenges to her own writing process and to her collaboration with Sullivan. Keller, struggling to write *Midstream,* a reflection on her "later life," characterizes her own writing process as disorganized and incoherent. Describing the fragmented typescripts she produced, she relates, "Into the tray of one's consciousness are tumbled thousands of scraps of experience. That tray holds you dismembered, so to speak. Your problem is to synthesize yourself and the world you also live in . . . into something like a coherent whole . . . I put together my pieces this way and that; but they will not dovetail properly" (*Midstream* 3). Sullivan and another collaborator assisted by sorting through the fragments, literally cutting them with scissors and pasting them together "into proper linear narrative, a narrative for consumption by readers from the seeing world" (Werner 973). This narrative was spelled into Keller's hands for her approval, but the spirit of the work remained dual-authored. As Marta Werner explains, "Despite the insistent use of the first-person narrator, *Midstream* is . . . a text written by two people though only one 'I' appears throughout" (982). Keller's "I," situated so intimately in touch, is never singular.

Touch in Question: Then and Now

Keller's audience also grappled with questions of her authorship, often exhibiting skepticism about her authenticity as a writer. She was first accused of plagiarism at the age of eleven, when she wrote a story, *The Frost King,* that was reprinted in the school's alumni magazine and then in a weekly publication for deaf and blind people, the *Goodson Gazette.* As the story gained wider readership, striking similarities emerged to a children's book by Margaret T. Canby. The editor published the matching phrasing and paragraphs, and an investigatory panel questioned Keller without Sullivan present. It was determined that Canby's story had been read to Keller years earlier, as she was just acquiring language, and she unintentionally reproduced elements of it. She was officially cleared of plagiarism but was traumatized by the events and subjected to skepticism of her abilities for decades. Kleege, dramatizing the events in a series of letters addressed to Helen Keller, calls the event an example of "consciousness on trial" and locates it as the "precise moment" when doubt entered Keller's mind regarding her own abilities (34). This trial, Kleege relates, questioned the basic elements of Keller's personhood. In *Blind Rage* she writes, "Face it, Helen. They may not say it out loud, but somewhere down deep they believe we are not quite human. Every time they repeated the question, 'How do you know what you know and remember what you remember?' they shoved you away from themselves. They drew a line in the sand and said, 'You don't belong to the same species. Being human means seeing and hearing.' . . . While they can imagine something they have never actually seen, they do not believe we can" (34–35).

As Kleege suggests, the charges of plagiarism and the general attitude of skepticism and disbelief concerning Keller's abilities are distancing tactics. By doubting and disbelieving Keller and her rhetorical productions, audiences separate themselves from her, drawing "a line in the sand" that prevents rhetorical identification between audience and rhetor. Kim Nielsen, in a biography of Sullivan, points out that not only was Keller's "capacity to generate original thought" on trial, but so was "Sullivan's legitimacy as a teacher" (117). If Keller and Sullivan's student-teacher relationship was based on lies, fraud, and manipulation—the reasoning went—then why should anybody believe that Keller was a rhetor with whom powerful rhetorical connections and identifications could be made?

The ownership of words between Sullivan and Keller was never completely clear in Keller's mind or in the minds of many in her audience. Keller felt that she and Sullivan were "pursued by misunderstandings" for the rest of their lives (*Midstream* 85). The editor of Keller's first book and the husband of Sullivan, John Macy, called their collaboration an "unanalyzeable kinship" (391). More recently scholars attempting to understand the relationship and its effects on writing have called the kinship a type of "collaborative consciousness" (Cressman 110), an example of "mirrored selves" (Swan "Touching Words" 343), or a process of "writing otherwise" (Werner). Keller herself struggled to articulate the ways in which she read and wrote as solitary or proprietary acts. At the age of twenty-eight she was once again suspected of plagiarism and responded to the charge by explaining how friends and acquaintances often "read"—spelled into her hands—"interesting fragments . . . in a promiscuous manner" (Lash 342). She admits that "it is not easy to trace the fugitive sentences and paragraphs" that have been spelled into her hand (343). She continues, "Sometimes I think I ought to stop writing altogether, since I cannot tell surely which of my ideas are borrowed feathers, except those which I gather from books raised in print" (343). With Braille, Keller could trace the origin of her ideas, but when words were spelled into her hands, they became part of her body, impossible to trace back or separate from herself. Her uncertainty over the ownership of her own ideas, a question paramount in her audiences' minds as well, caused her to consider ceasing to write altogether. Ending *Midstream,* she relates, regretfully, "I have written the last line of the last autobiography I shall write . . . I lift my tired hands from the typewriter. I am free" (Keller *Midstream* 342). With a sense of failure, she assesses, "My autobiography is not a great work" (343).

Even after more than a century following the charges of plagiarism concerning Keller, scholars continue to wrestle with the questions surrounding Keller and Sullivan's specific partnership; more broadly, touch is still questioned in rhetorical productions made by people with disabilities today, especially those who use forms of facilitated or mediated communication. It is little wonder that Keller or other rhetors who use touch to communicate today continue to be doubted by themselves and by their audiences because few, if any, existing rhetorical models are available for touch. With the rise of collaborative writing models and writing practices situated in participatory media and new media contexts, the rhetor has been imagined as plural and writing is generally accepted as a social act that takes place in a context that is shared among many people.[3] These challenges and revisions, however, do not fully account for the experience of disability in rhetorical production, especially when disabled rhetors depend on touch to communicate and work closely with a facilitator or someone else to construct a message. As Lisa Ede and Andrea Lunsford point out, "We are often more comfortable theorizing about subjectivity, agency, and authorship than we are attempting to enact alternatives to conventional assumptions and practices" (356). Since the publication of their collaboratively authored text in the mid-1980s, they assert, "we have been calling on scholars in rhetoric and composition, and in the humanities more generally, to *enact* contemporary critiques of the author and of the autonomous individual through a greater interest in and adoption of collaborative writing practices—and to do so not only in classrooms but in scholarly and professional work as well" (355–56). They remark that although there have been some efforts to respond to their call, "in general we would have to characterize these responses as limited" (356). In the context of disability, this response is particularly limited, especially in relation to ways in which collaborative writing and communication are enacted.

To this day disbelief continues to circulate when touch is used rhetorically, particularly in contexts of disability. The controversy over the technique of facilitated communication (FC), a method of communicative support for people with disabilities such as autism, demonstrates this continued suspicion of touch in rhetorical production most clearly. With FC, a facilitator provides hand, wrist, elbow, or arm support to a person with a communicative disability, so that he or she can point to letters on a letter or picture board or press keys on a keypad, computer, or other type of assistive communication device. As Cynthia Lewiecki-Wilson explains, FC provokes anxiety in cultural conversation and disciplinary categories "because it clashes with the liberal model of a core, autonomous, and stable self and the tropes of validity and supervision that as Michael Warner argues, accompanied and comprised the construct of the universal and abstract citizen-subject that arose in the eighteenth century along with the epistemology of science" ("Rethinking Rhetoric" 160). In short, touch calls into question the assumptions of the liberal subject.

Debates about FC, especially about the authenticity of the messages of FC users, are rooted in suspicion, fear, and doubt about how touch circulates rhetorically and about the dangers of touch for uncommunicative or semicommunicative users. As Ralph James Savarese explains, FC conversations in the early 1990s "became entangled in the sex abuse hysteria of the same period, with children typing allegations against their parents and caregivers, and in unsympathetic authentication procedures that revealed questionable authorship" (xxi). Some of the allegations of fraudulent communication were true, and charges of sexual abuse are *always* serious, but FC, which the neurologist Margaret Baumann says was "oversold in the beginning," became like "the proverbial baby, 'thrown out with the bath water'" (xxiv). In a *Frontline* documentary, the technique was compared to the Ouija board effect, whereby the facilitator, consciously or unconsciously, moves the user's hands; allegations of hoaxing and quackery ensued. As Savarese explains, "It became too easy to suspect a method that often had, at least at the outset, a facilitator's hand entwined in the Autist's" and was "dismissed outright" (xxi). Even though proponents of FC such as Biklen have continued their research with successful results and some users of FC have achieved independent typing skills, proving in some cases beyond a doubt that they are the authors of their messages, FC continues to be doubted. Savarese identifies a "passion to *disbelieve*" in continuing discussions about FC, even as, possibly, the "tide is beginning to turn," following attention recently paid to FC users such as Sue Rubin, Jamie Burke, and Sean Sokler (xxii–xxiii).

Similar to how Keller's consciousness was on trial in the wake of plagiarism charges against her, the users of FC and other forms of assisted, tactile, or mediated communication face questions regarding their status as thinking, feeling, and expressive beings in the continuing debates regarding augmentative communication today. As Savarese suggests, FC troubled the "theory of mind" approach to classical autism predominant at the time, disrupting the dominant idea that people with autism or other disabilities cannot imagine other mental states. The controversy surrounding the messages of autistics who use FC is a symptom of a larger tendency to doubt and devalue the rhetorical productions of disabled rhetors. As G. Thomas Couser has explored in *Vulnerable Subjects,* collaborative life writing for people with a range of different disabilities is fraught with complex ethical and representational questions. At the heart of continuing questions regarding the communicative abilities and potentials of people with a wide range of disabilities are the entangled issues of belief, identification, and connection. When a nonnormative rhetor is dismissed as inhuman, unconscious, and unfeeling; doubted regarding her abilities; or accused of plagiarism, rhetorical identification is nearly impossible. When a facilitator or assistant is implicated in a scandal or accused of lying and fraud, the partnership between rhetor and facilitator becomes unbelievable and unworthy of study or understanding. Inquest panels, investigative reports, and court cases become the focus, when efforts to understand touch as rhetorical are most needed.

How is touch rhetorical? To begin to answer this question, in this chapter I track the relationship between ability and independence in the rhetorical tradition's preference for a singular rhetor. Despite this assumption of independence, I identify opportunities for supporting rhetorical touch among interdependent rhetors by rereading Aristotle's categorization of rhetoric as a *dynamis,* an art based on potential, that can be understood, based on Aristotle's other definitions of art and sensory perception, to foster alternative forms of rhetoric among disabled rhetors. To prod this potential of rhetoric to accommodate multiple, interdependent rhetors and nondiscursive expressions such as touch, I employ Kenneth Burke's concept of rhetorical identification, exploring it through his invocation of the philosopher Empedocles. I read Keller and Sullivan's rhetoric through a model attentive to touch, contrasting it to the limits of existing rhetorical situations, and show how identification among multiple rhetors and audiences can be facilitated through an understanding of touch as rhetorical. Then I posit a new model of rhetoric that values touch as rhetorical by imbuing the traditional triads of rhetoric—*ethos/pathos/logos* and speaker-writer/audience/message—with redefinitions of three key terms—felt *logos, mētis,* and *kairos*—which I consider in-depth in successive chapters. In this model, touch is rhetorical in that it communicates a message, feeling, or connection among and between multiple bodies, responding to physical, situational, and rhetorical stimuli and shaping potential spaces for identification among a variety of bodies, not in spite of but because of communication differences and bodily diversities.

The Rhetoric, the Rhetor, and the Potentials of *Dynamis*

Aristotle's definition of rhetoric, at first blush, seems particularly unaccommodating to disabled rhetors. Ability is integral to Aristotle's definition: "Let rhetoric be [defined as] an ability, in each [particular] case, to see the available means of persuasion" (*On Rhetoric* 1355b1; trans. Kennedy 36). Ability is built into this definition of "rhetoric," requiring the rhetor to possess the abilities to understand, to evaluate, and to apprehend. Independence is closely connected to this assumption of ability. Aristotle, noting the differences between the rhetor and the sophist, states that "in the case of rhetoric . . . there is the difference that one person will be [called] rhetor on the basis of his knowledge and another on the basis of his deliberate choice, while in dialectic *sophist* refers to deliberate choice [of specious arguments], *dialectician* not to deliberate choice, but to ability [at argument generally]" (1355b14; trans. Kennedy 35–36). The rhetor is singular, "one person," who exercises his ability to share his knowledge. Aristotle associates ability in rhetoric with truth, knowledge, and independence, leaving little possibility for disabled or interdependent rhetors. As Cynthia Lewiecki-Wilson explains, "rhetoric's received tradition of emphasis on the individual rhetor" is one that is particularly problematic for disabled rhetors who may work with aids or assistants to communicate, because it "confirms the existence of a fixed, core self, imagined to be located in the mind" ("Rethinking Rhetoric" 157).

The *pisteis,* often considered the core of Aristotle's rhetoric, operating as means of persuasion, implement this ability. Aristotle states, "Of the *pisteis* provided through speech there are three species: for some are in the character [*ēthos*] of the speaker, and some in disposing the listener in some way, and some in the argument [*logos*] itself, by showing or seeming to show something" (*On Rhetoric* 1356a3; trans. Kennedy 37). The proofs are *entechnic,* "embodied in art," "intrinsic," and invented (trans. Kennedy 37). Extrapolated from Aristotle's definition of rhetoric aligned with ability, knowledge, and independence, the appeals seem rooted in assumptions unfriendly to disabled rhetors: character that is independent and able, audiences who are capable of certain dispositions, and arguments that are based in an understanding of truth and knowledge rooted in ability.

Inconsistencies, however, in Aristotle's delineation of the *pisteis* paired with closer examinations of his theories on art and sense perception reveal possibilities for alternative understandings of the artistic proofs beyond the preference of an independent, able rhetor and toward potentials for interdependent, diverse rhetors who use touch rhetorically. Kennedy, stating that "the most problematic of Aristotle's rhetorical terms are *pistis* . . . *ēthos,* [and] *pathos,*" notes that there are at least three different meanings of "proof" that circulate in Aristotle's text (*A New History* 60).[4] These multiple and different meanings invite different interpretations based on different contexts. Aristotle, for example, may have conceived of other means of persuasion or *pisteis* beyond speech. Kennedy explains that Aristotle's definition "assumes but does not specify that rhetoric is a quality of speech, since the root of the word (*rhē-*) refers to speaking" (57).

Significantly, Aristotle "identifies the genus to which rhetoric belongs as *dynamis:* 'ability, capacity, faculty'" (Kennedy *On Rhetoric* 36n34). This categorization suggests multiple potentials for rhetoric and raises the possibility that Aristotle recognized the dynamic potentials of rhetoric produced by alternative means, including those of touch. The element of ability in Aristotle's definition of rhetoric, when paired with the potential and capacity of *dynamis,* is more inviting to non-normative rhetors, especially those who may use touch rhetorically. Aristotle's definitions of art in his other works also reveal potentials for rhetoric beyond the realm of the verbal and into the haptic realm. In his *Nicomachean Ethics,* Aristotle defines art as a "coming into being," a definition which shares key elements with his definition of rhetoric. According to Aristotle, "All art is concerned with coming into being, i.e. with contriving and considering how something may come into being which is capable of either being or not being, and whose origin is in the maker and not in the thing made" (1140a; trans. Ross and ed. Brown 105). As Kennedy relates, this definition means that "art is thus for [Aristotle] not the product of artistic skill, but the skill itself" (*On Rhetoric* 36n34). This characterization of art as "coming into being" aligns with Aristotle's notion of rhetoric as an art belonging to the genus of *dynamis,* a categorization less focused on product than on potential and capacity. In Aristotle's philosophical writing, "*dynamis* is the regular word for 'potentiality' in a matter or form that is 'actualized' by an efficient cause" in which the "actuality produced by the potentiality of rhetoric is not the written or oral text of a speech, or even persuasion but the art of 'seeing' how persuasion may be effected" (36n34). Rhetoric, like other arts, is produced by a transformative "coming into being" and is based on potential for "seeing" how persuasion works in multiple ways. The *pisteis,* which are "embodied in art" in Aristotle's rhetoric, possess similar qualities of transformation and change. Aristotle emphasizes the process and potential rather than the products of art and rhetoric, highlighting the possibility for transformation and change in both pursuits.

Aristotle's definitions of art and rhetoric also share an important characteristic with his definitions of sensory perception. In his intensive study of touch in *De Anima,* Aristotle characterizes touch as a process of change and transformation similar to the process of "coming into being" in art and the *dynamis* of rhetoric. In *De Anima,* Aristotle classifies all the senses, including touch, as *dunameis,* or capacities, and defines the senses through actualizations (*energeiai*) of these potentials (Freeland 228).[5] As is the case with rhetoric and art, which are also capacities or potentialities, the senses are not simply products—senses in and of themselves—but become senses through an actualization or transformative process. Perception through the senses, including touch, is an active process by which an organism is "moved and affected" by an alteration (*alloiosis*) of the sense faculty in which perception is both potential and actual (*De Anima* 416b-417a; trans. Foster and Humphries 234). Aristotle characterizes this alteration of sense perception through the process of *aesthesis,* which is the "faculty by means of which we are able to characterize or identify things as a result of the use of our senses" (Hamlyn 6). As Mark Paterson explains, elsewhere in book 2 of *De Anima,* Aristotle also describes *aesthesis* as a process of change using the tactilely inflected word *paschein.* The literal translation of *paschein* reveals an "ambiguity of feeling," containing both the alteration (*alloiosis*) of touch through a physical sensory faculty and the psychological phenomena of tactile perception (*haptesthai*) (Paterson 19). Furthermore *paschein,* which can also imply suffering, conveys an active undergoing rather than the passive experience of pain, a kind of *pathos* (Paterson 19).[6] For Aristotle, these terminological associations suggest that touch functions as a potential in a particularly responsive and dynamic register of perception.

Like Aristotle's approaches to rhetoric and art, the sensory perception of touch is not a product but instead is the process by which change and transformation are possible. Touch is a coming into being, like art or rhetoric, that depends on the potential of bodies coming into contact. Perception, for Aristotle, "is not the activity of the objects themselves, then, but the way the sense faculty (*aesthesis*) is affected or altered; and hence the sensory faculty allows both actual and potential activity, and furthermore has the capacity to be receptive" (Paterson 19). Similar to rhetoric, sensory perception is not perception in and of itself but is the transformative and receptive process by which change happens. Sensory perception is not simply a product or activity but the way that transformation or alteration happens or potentially happens; it is a *dynamis.*

This process of sensing is ripe with potential for the art of rhetoric beyond verbal or visual means. The potential of *dynamis* in rhetoric—the art of "seeing" the available means of persuasion—possesses possibilities even beyond verbal or visual apprehension and toward the opportunities afforded by touch. As Kennedy relates, "to see" in Aristotle's definition of rhetoric "translates [as] *theorēsai,* to be an observer of and to grasp the meaning or utility of" and is related to "theory" (*theoria*), which literally means "to see," demonstrating the "visual imagery common in the *Rhetoric*" (*On Rhetoric* 37n34, 29n5). Beyond this visual imagery, however, Aristotle suggests that meaning can be *grasped,* imbuing rhetoric with a potentially tactile element. This act of "grasp[ing] the meaning or utility of" is a more active and transformative role than one of "observer." This active process of grasping meaning is well aligned with the aims of art—to actualize a process of change or "coming into being." In fact Aristotle evokes the act of grasping when he describes the *pisteis* again: "Since *pisteis* come about through these [three means], it is clear that to grasp an understanding of them is the function of one who can form syllogisms and be observant about characters and virtues and, third, about emotions" (39). This act of grasping meaning suggests multisensory avenues for rhetoric and inclusive means of persuasion beyond verbal or visual apprehension. For Aristotle, the potentials of rhetoric as an art rest in the act of receptive and perceptive grasping—sensing an understanding about a rhetorical context or situation and potentially actualizing a response to it. The *pisteis,* means of persuasion "embodied in art" and associated with sensory perception and the tactile act of grasping, suggest possibilities for disabled rhetors who use touch rhetorically.

For people with disabilities, Aristotle's definition of rhetoric based on ability and the independent rhetor can seem normative and limiting; yet his classification of rhetoric as a *dynamis* emphasizes capability, potentiality, and connection. The *dynamis* of rhetoric shares the *dynamis* of sensory perception in Aristotle's philosophy, opening up the element of "ability" in Aristotle's definition of rhetoric to a wider range of possibilities. Understood in conjunction with his theories of art and perception, Aristotle's rhetoric reveals itself to be aligned with a transformative and dynamic process of sensing, based on potential and capacity. This rhetoric based on potential and sensing in multiple ways provides possibilities for rhetors with disabilities. As Lewiecki-Wilson, drawing from Aristotle's classification of rhetoric as a *dynamis,* suggests in her theorization of rhetoric for people with mental disabilities, "We need an expanded understanding of rhetoricity as potential, and a broadened concept of rhetoric to include collaborative and mediated rhetorics that work with the performative rhetoric of bodies that 'speak' with/out language" ("Rethinking Rhetoric" 157). In effect, rhetoric is a potential, an art of "seeing," feeling, or sensing how persuasion can work, and rhetors who use their bodies to communicate in nontraditional ways tap into this potential with and without language. This transformative potential of rhetoric's *dynamis,* especially understood in relation to Aristotle's complementary theories of art and sensory perception, reveals possibilities for nonnormative or disabled rhetors who may use a range of rhetorical strategies, beyond spoken and written language, to communicate. Touch is a *dynamis* that brings bodies into being together, outside of normal channels of communication. People with disabilities who use touch as a rhetorical strategy, such as Keller, and others, such as those who use FC, are doing so in the spirit of Aristotle's classification of rhetoric as a *dynamis.*

Recognizing touch as fitting into rhetoric's classification as a *dynamis* also supports understanding rhetorical production beyond the preference of a singular, independent rhetor. Touch, as it functions in both perception and reception between bodies, is a "coming into being," an active process of potentiality and actuality among bodies. Touch comes into being when a body makes physical contact with itself or another body, creating a new sensation and alternative configuration of stimulus and response, both physical and psychological. A touch that is rhetorical is not necessarily the product of a sensation or perception but instead is the process of transformation initiated by touch. In other words, rhetorical touch is a potentiality and a *dynamis,* an active coming into being among bodies. Rhetorical touch, like rhetoric itself, responds to specific situations and crafts specific meanings contingent to a particular time, place, and culture. Touch operates as a rhetorical art, bringing bodies together in dynamic potentials.

The Limits and Potentials of Rhetorical Situations

Although the *pisteis* demonstrate potential for supporting rhetors who use nonnormative rhetorical strategies such as those based on touch, contemporary understandings of Aristotle's treatment of the *pisteis* do not fully realize the potential for haptic rhetorics that span multiple bodies of diverse abilities. Kennedy points out that Aristotle does not use terms such as *ethos* and *pathos* in "the technical sense" such as "rhetorical *ethos*" but instead typically to refer to "an attribute of persons, not of a speech" (*On Rhetoric* 37n40). Consequently the "shorthand ethos-pathos-logos to describe the modes of persuasion is a convenience that does not represent Aristotle's own usage" (37n40–38n40). This misunderstanding may be rooted in Aristotle's own simplification in his statement that a "speech [situation] consists of three things: a speaker and a subject on which he speaks and someone addressed" (*On Rhetoric* 1358a1; trans. Kennedy 47). This triad of elements—speaker, message, audience—is a long-standing area of continued debate, inquiry, and revision among rhetorical theorists and compositionists. Charles Ogden and I. A. Richards, James Kinneavy, and Lloyd F. Bitzer all use a triad to describe the elements of rhetoric.[7] Those who respond to Bitzer, including Richard Vatz, Scott Consigny, and Barbara Biesecker, among others, continue to work with and revise the triad of elements for contemporary needs.

Bitzer, Vatz, and Consigny emphasize different elements of the triad but share a preference for a rhetoric based on ability, emphasizing agential rhetors and capable audiences. Bitzer assumes able audience members who are "capable" and can act as "mediators of change" (8). In his featuring of the "phenomenological perspective of the speaker" or "interpreter," Vatz mainly assumes a rhetor who is independent, autonomous, and agential, who adeptly makes choices regarding salience of information and translates that information accordingly (154). Consigny's proposed resolution to the debate between Bitzer's focus on the situation and Vatz's focus on the speaker—developing the art of topic selection—also emphasizes the agency and independence of "the rhetor," who possesses "command," "control," and "mastery of topics" ("Rhetoric and Its Situations" 184–85). Furthermore, Bitzer, Vatz, and Consigny each privilege discourse as the primary type of rhetoric suited for responding to exigence or creating reality. Bitzer appears to suggest that the rhetorical situation may be composed of nonverbal elements such as a "complex" involving persons, events, objects, and relations but limits this potentiality by focusing on discourse as the sole modification of exigence (6). Vatz is primarily interested in the "linguistic depiction" of a situation created by a rhetor, such as a politician or editor (157). The topics in Consigny's approach mainly rely on verbal and discursive examples ("Rhetoric and Its Situations" 183), although topics can function as "places as perception, discovery and explanation of the unknown" (McKeon 205). In all of these approaches, rhetoric is primarily assumed to be a spoken or written art, made by an independent, able, and autonomous rhetor and audience.

More recent reformulations of the rhetorical situation begin to open up productive possibilities for reimagining rhetoric as a *dynamis* arising among multiple relationships across various interdependent agents and registers. Biesecker critiques the assumptions that drive theorizations of the rhetorical situation, especially the "assumption that a logic of influence structures the relations" of the rhetorical situation and the related assumption of speaker and audience "identity [as] constituted in a terrain different from and external to the particular rhetorical situation" ("Rethinking the Rhetorical Situation" 110). David Hunsaker and Craig Smith suggest a more nuanced understanding of audience, offering categorizations such as situational audience, rhetorical audience, and actual audience. Smith and Scott Lybarger extend this notion of multiple audiences into a theory of the "multiplicity of discourses" to account for "postmodern multiplicities" in methods, media, perceptions, and exigencies (199, 197). Mary Garret and Xiaosui Xiao too recognize multiple audiences and particularities, arguing for "placing much greater stress on the interactive, organic nature of the rhetorical situation" (31).[8] Approaches such as these conjure the possibilities of Aristotle's classification of rhetoric as a *dynamis,* full of potential for nonnormative and interdependent rhetors.

Other rhetorical theorists too have moved toward delimiting rhetorical situations, suggesting new spaces for analysis that dethrone any single element of the rhetorical situation and invite nondiscursive possibilities in the spirit of the most transformative possibilities of Aristotle's categorization of rhetoric as an art based on *dynamis*. Jenny Edbauer has extended the concept of rhetorical situations to the notion of rhetorical ecologies, especially in relation to responses to public places that include both "material experiences and public feelings" (5). Edbauer describes elements of rhetoric beyond the triad of speaker, audience, and message, adding places, affect, events, and enactments, which operate in an ongoing flux in relation to space and time: the "*elements of rhetorical situation simply bleed*" into and through each other in rhetorical ecologies that include Raymond Williams's concept of "shared structures of feeling" (9, 20). Similarly theorists such as Margaret Syverson, Nedra Reynolds, and Louise Wetherbee Phelps have described rhetoric and writing as distributed across multiple people, places, bodies, and social structures. Syverson notes that the rhetorical triangle "has tended to single out the writer, the text, or the audience as the focus of analysis" (23). Far from this static triangularized notion of writing, she argues for recognizing composition as distributed "across physical, social, psychological, and temporal dimensions. . . . the social dimensions of composition are distributed, embodied, emergent, and enactive" (23). This attention to dimensions of feeling and to the physical and embodied elements of composition signals a ripening of the rhetorical situation and a readiness to explore touch as rhetoric operating across shared structures of feeling and physical and psychological dimensions.

Although the sense of touch has not been considered comprehensively in rhetorical theory, recognizing touch as rhetorical furthers this development of the rhetorical situation into more capacious understandings of rhetoric in relation to other bodies, entities, places, enactments, and various other elements. The turn toward multiple audiences and speakers in refigurations of the rhetorical situation such as those by Ede and Lunsford ("Among the Audience") is also promising for people of wide ranges of ability who produce rhetoric in relation to and with the physical and emotional support of others. Rhetorical situations such as these are embodied, distributed, affective, and informed by social and material forces. To undertake more fully the transition from the rhetorical situation to more transformative models, I turn to Burke's theory of rhetorical identification, a theory that by definition brings bodies together in physical, sensory, and discursive contact. In identification "the rhetor," "the audience," and "the message" become mixed and indeterminate. Each element is dependent on another, resulting in a rhetoric of sensation that fulfills the potentials of nonverbal rhetorics suggested by Aristotle but not taken up comprehensively in the importation of the elements of rhetoric in contemporary theory. For rhetors who produce rhetoric in physical relation to other bodies, identification by sensation responds to the complexity and dynamics of their rhetorical situations.

Burkean Rhetorical Identifications

Burkean identification unfolds the potential of rhetoric as a *dynamis,* bringing together bodies and nondiscursive ways of relating. Burke's treatment of traditional rhetoric, especially Aristotle's *Rhetoric,* shows that he is interested in forming new, interdependent relationships between speaker, audience, and message. In *A Rhetoric of Motives,* Burke begins staking out new relationships among the elements of rhetoric by focusing on the concept of audience. He transitions from the key term of ancient rhetoric, "persuasion," to his new term for modern rhetoric, "identification." This transition requires blurring the lines between "internal" and "external" elements of rhetoric, including the appeals and the roles of audience and rhetor.

Burke classifies the appeals of traditional rhetoric as operating under an assumption that the audience is external. He explains that in "traditional Rhetoric, the relation to an external audience is stressed," noting that "Aristotle's *Art of Rhetoric,* for instance, deals with the appeal to audiences in this primary sense: It lists typical beliefs, so that the speaker may choose among them the ones with which he would favorably identify his cause or unfavorably identify the cause of an opponent; and it lists the traits of character with which the speaker should seek to identify himself, as a way of disposing an audience favorably towards him" (*A Rhetoric of Motives* 38). Burke updates this traditional triad of *logos-ethos-pathos* and message-speaker-audience, however, for "modern" rhetoric, which "must also concern itself with the thought that, under the heading of appeal to audiences, would also be included any ideas or images privately addressed to the individual self. . . . For you become your own audience . . . when you become involved in psychologically stylistic subterfuges for presenting your own case to yourself in sympathetic terms" (38–39). By making the rhetor part of the audience, Burke combines internal and external elements of rhetoric. Whereas ancient rhetoric stresses the "external" audience, modern rhetoric blurs this distinction by internalizing appeals to the audience within the self. Internal or "intrinsic" *entechnic* appeals that are "embodied in art" are also embodied in both the rhetor and the audience, becoming external. The line between the rhetor's body and the audience's bodies is porous in this reformulation.

Burke furthers this blurring between internal and external elements of rhetoric by explicitly including nonverbal expressions such as images and ideas, including physical and tactile elements, in his notion of how internal and external pressures and counterpressures shape identification. "The individual person, striving to form himself in accordance with the communicative norms that match the cooperative ways of his society, is by the same token concerned with the rhetoric of identification. To act upon himself persuasively, he must variously resort to images and ideas that are formative. Education ('indoctrination') exerts such pressure upon him from without; he completes the process from within. If he does not somehow act to tell himself (his own audience) what the various brands of rhetorician have told him, his persuasion is not complete" (*A Rhetoric of Motives* 39). Again, Burke merges the rhetor with the audience, matching internal with external pressures in the process of identification via the individual person with society. Rhetorical identification is contingent on the parallel exertions of pressure from within and without. Burke also recognizes the importance of "communicative norms" that direct identification in a society and seeks to adjust traditional rhetoric, especially Aristotelian rhetoric, for the needs of a modern culture. He notes that "since 'rhetoric,' 'oratory,' and 'eloquence' all come from roots meaning 'to speak,' you can have the Aristotelian stress upon rhetoric as *sheer words*" but that other emphases are possible (51). Like Kennedy, Burke recognizes that the root of "rhetoric" (*rhe*) refers to speaking in Aristotelian rhetoric, but he also sees potential beyond words.

Burke characterizes Aristotle's rhetoric in agonistic and physical terms, writing that Aristotle "gives what amounts to a handbook on a manly art of self-defense," including "the holds and the counter-holds, the blows and the ways of blocking them, for every means of persuasion the corresponding means of dissuasion, for every proof and disproof" (*A Rhetoric of Motives* 52). In response Burke offers a more subtle but no less physical art of rhetoric, such as the matching of internal and external forces, exertions of pressures and counterpressures, and a variety of nonverbal maneuvers beyond "sheer words." He urges the rhetorician to participate in "analysis of non-verbal factors wholly extrinsic to the rhetorical expression considered purely as a verbal structure" and recognizes that rhetoric can work by inducing "an act beyond the area of verbal expression considered in and for itself" (62, 64). Furthermore, Burke advises "recognition of nonverbal, situational factors that can participate" in the effectiveness of rhetoric, particularly through Bronislas Malinowski's idea of the "context of situation" (65). Burke's formulation of a rhetorical situation is imbued with nonverbal elements of rhetoric, including specific attention to the forces of pressures, both internal and external. Nonverbal and tactile elements are integral to his movement from traditional rhetorics of persuasion to rhetorics based on identification. Burke's identification catalyzes Aristotle's classification of rhetoric as a *dynamis,* exploring the potentials of rhetoric beyond verbal means and toward rhetorics of sensation.

Burkean identification is explicitly and implicitly tactile. The process of identification always starts with at least two bodies. Burke uses the example of "Colleague A" and "Colleague B," but any two bodies suffice. Rather than the singularity and independence of "the rhetor" or "the speaker" assumed in ancient rhetoric and often imported into contemporary retheorizations of traditional rhetoric, Burkean identification, starting with two bodies, also includes a "doctrine of *consubstantiality,* either explicit or implicit," that includes the merging of "substances," which Burke defines in explicitly tactile terms (*A Rhetoric of Motives* 21). Specifically, Burke identifies the substances of identification that make two or more people "consubstantial" as existing in "common sensations, concepts, images, ideas, [and] attitudes," which induce ways of "*acting-together*" among people (21). Although he does not explicitly identify touch as a specific "common sensation," touch possesses the potential to qualify as a particularly transformative process of identification in Burkean terms. Touch is an "*acting-together*" that propels identification among bodies in contact. This "*acting-together,*" similar to Aristotle's "coming into being," brings bodies together in a process of change and alteration.

Identification is also implicitly tactile, sharing key characteristics and features with the sense of touch. Burke uses identification as "an instrument . . . to mark off the areas of rhetoric," going "beyond the traditional bounds of rhetoric" toward an "intermediate area of expression that is not wholly deliberate, yet not wholly unconscious," one that "lies midway between aimless utterance and speech directly purposive" (*A Rhetoric of Motives* xiii). Burke similarly uses identification to discover "how a rhetorical motive is often present where it is not usually recognized, or thought to belong" (xiii). This characterization of identification implicitly conjures the qualities of tactility and physicality that Burke first introduces with his blurring of internal and external elements of rhetoric and his notion of pressure and counterpressure shaping identification. Touch is often experienced as an "intermediate area of expression" between two bodies, existing in ambiguous relation to "utterance" and "speech." Touch, often not completely "deliberate" or completely "unconscious," reveals a wide range of rhetorical meaning in areas in which "it is not usually recognized, or thought to belong."

In addition touch can be understood to qualify, in Burke's terms, as an invitation to rhetoric. As Burke relates, "Put identification and division ambiguously together, so that you cannot know for certain just where one ends and the other begins, and you have the characteristic invitation to rhetoric" (*A Rhetoric of Motives* 25). Touch is also a sensation in which, as with identification and division, one may not know for sure where one body or feeling ends and another begins. As numerous theorists of touch such as Maurice Merleau-Ponty and others have described, when two bodies come into contact, it is difficult to discern what part is the touched or the toucher, or even where exactly one body ends and the other begins. Touch embodies the blurring of internal and external rhetorical elements and the tension of pressure and counterpressure key to Burke's characterizations of identification and the invitation to rhetoric.

In *A Grammar of Motives,* Burke provides more insight into how bodies in contact form rhetorical identification, offering a context for understanding his listing of common sensations as a vehicle for identification in *A Rhetoric of Motives.* Exploring the paradoxical key term "substance," Burke considers the relationship of intrinsic and extrinsic bodies via theories of the skin. He writes, "We can take it as a reliable rule of thumb that, whenever we find a distinction between the internal and the external, the intrinsic and the extrinsic, the within and the without, (as with Korzybski's distinction between happenings 'inside the skin' and happenings 'outside the skin') we can expect to encounter the paradoxes of substance" (*A Grammar of Motives* 47). Alfred Korzybski recognizes the paradox of both a correspondence and a disconnection between the inside and the outside of the skin: "The events outside our skin are neither cold nor warm, green nor red, sweet nor bitter, but these characteristics are manufactured by our nervous system inside our skins, as responses only to different energy manifestations, physiochemical processes" (*Science and Sanity* 384). The sense of touch also characterizes these paradoxes of substance because touch, like substance, is both material and symbolic and exists as internal and external experience. This paradox of skin informs the structure of bodies for Burke: "Since the body is but chemistry, and all outside the body is but chemistry, the very mode of thought that forms a concept of the 'intrinsic' in these terms must also by the same terms dissolve it" (*A Grammar of Motives* 47). Sensations felt on the skin connect internal and external areas, rendering each nearly indivisible and, for an instant, both connected and separate. Registering sensation inside the skin also means registering sensation outside the skin, dissolving categories of intrinsic or extrinsic. The paradoxes of skin, touch, and sensation exemplify the characteristics of Burkean identification—as Colleagues A and B merge, so do the insides and outsides of substance; as Colleagues A and B divide, so do the inside and outside of their worlds, on material and symbolic levels.

Identification through consubstantiality forwards Aristotle's classification of rhetoric as a *dynamis,* encouraging potentials for transformation and change among participants. Burke invokes substance in this capacity in *A Rhetoric of Motives,* writing that when someone is identified with someone else, "he is both joined and separate, at once a distinct substance and consubstantial with another" (21). This transformation is an intrinsic and extrinsic process among at least two people. As Burke notes, "If men were wholly and truly of one substance, absolute communication would be of man's very essence" (22). Instead rhetorical identification takes on these paradoxes of substance, exemplified in the inside and outside of skin and bodies, to "lead us through the Scramble, the Wrangle of the Market Place, the flurries and flare-ups of the Human Barnyard, the Give and Take [and] the wavering line of pressure and counterpressure" (23). In this characterization

Burke repeats the image of internal and external pressure and counterpressure, using the concept of a "wavering line" between two people and their interests, noting that "rival rhetoricians can draw [this line] at different places, and their persuasiveness varies with the resources each has at his command" (25). This "wavering line" of pressure and counterpressure, internal and external, connection and division means that "we must think of rhetoric not in terms of some one particular address, but as a general *body of identifications*" (26). This body of identifications has the potential to be physical as much as it is discursive and rhetorical, responding to internal and external pressures and comprising itself of bodies rather than a singular, independent body.

Even more specifically, touch has the capacity to reveal the *bodies* of identification involved in verbal and nonverbal rhetoric. Touch is "pressure" and "counterpressure," a "wavering line" between inside and out, a substance both of distinction and consubstantiality, and a location for forming identifications as well as divisions among bodies in contact. For Burke, identification and consubstantiality among bodies produce cooperative actions, which often take the shape of linguistic forms of rhetoric but also include a wider array of nondiscursive acts such as ideas, attitudes, images, and sensations. This potential prods traditional rhetoric, especially Aristotelian rhetoric, to consider more comprehensively persuasion beyond explicit verbal and linguistic means. As Burke remarks, "Aristotle's Rhetoric centers in the speaker's explicit designs with regard to the confronting of an audience. But there are also ways in which we *spontaneously, intuitively, even unconsciously* persuade ourselves" (*Language as Symbolic Action* 301). In these ways rhetoric reaches beyond the verbal realm, into the "nonverbal, postverbal, and superverbal" (*A Rhetoric of Motives* 180). For Burke, this "jumping-off place" beyond the verbal and toward the "visceral" and "unutterable" gives rise to an "inclusive nature [that] would be more-than-verbal rather than less-than-verbal" (*A Rhetoric of Motives* 180). Rhetoric functions in the visceral locations that touch creates between more and less verbal, often working to facilitate identifications spontaneously, intuitively, and unconsciously, frequently including diverse bodies rather than excluding them.

Identification, Proof, and the Bodies of Touch

To exemplify identification, Burke mobilizes images and themes involving bodies of diverse abilities and disabilities coming together in contact and movement. He contrasts the imagery of suicide and death in John Milton's "Samson Agonistes" with Matthew Arnold's poem "Empedocles on Etna," focusing particularly on the bodies of each character and author. Burke is interested in "the correspondence between Milton's blindness and Samson's" and notes both individual and factional identification between multiple bodies, including Milton, Samson, and God (*A*

Rhetoric of Motives 4). Forwarding the doctrine of consubstantiality, he explains, "In saying, with fervor, that a blind Biblical hero *did* conquer, the poet is 'substantially' saying that he in his blindness *will* conquer" (5). He suggests that by identifying with the self-destructive Samson, Milton was able to explore his own "suicidal motive" even as his religion strictly forbade suicide.

Burke compares Milton's identification with Samson to Arnold's poem recounting Empedocles's self-immolation at Etna. Empedocles throws himself into the volcano to achieve unification or "oneness" with the cosmos, an act that also seeks to exemplify and prove his theories regarding the origins of life. Empedocles's basic worldview, an oscillation between sameness and difference and oneness and multiplicity, is typically known as a theory of pores and effluences. Moved by the two opposing but complementary forces of Love and Strife, four basic elements—air, water, fire, and earth—intermingle and connect via pores and effluences, forming the root of everything from bodies and matter to words and song. According to Burke, "Empedocles sees promise of freedom as he 'plunges into the center,' a self-immolation that unites him idealistically with mountain, sea, stars, and air, while the volcano, into which he had hurled himself, is also by legend the prison of a buried titan" (*A Rhetoric of Motives* 7). Similarly to how Milton's and Samson's blindness are "substantially" alike, Empedocles's leap into the volcano shows him attempting to become substantially one with the cosmos and the elements he believed comprised the world. At the level of substance, Empedocles's body—his pores and effluences—intermingles so completely with the pores and effluences of the natural environment that they become one, via shared sensation. Burke notes similar motives in Arnold's poem "Sohrab and Rustum," the story of a son killed unknowingly by his father in combat. Like Empedocles, whom Burke calls the "mystic" who sought "devout identification with the source of all being," Rustum realizes his son's identity as he dies and seeks similar identification: "Through the progress of the river to the sea he plunges by proxy into the universal home" (xiv, 8). The "same cosmically unifying end" (8), an end typified by bodies in motion, becoming completely one with and identified with other natural bodies, characterizes both Empedocles's jump into the volcano and Rustum's passage out to sea.[9]

Burke further connects Empedocles's and Rustum's actions by focusing on their bodies as exemplifying the principle of transformation. Interested in finding "*a motive that can serve as ground for both these choices,* a motive that, while not being exactly either one or the other, can ambiguously contain them both," Burke posits that the image of transformation contains both motives, operating as a principle that transcends various differences dialectically (*A Rhetoric of Motives* 10). For Burke, "killing, being killed, and the killing of the self might all localize the same principle of transformation" (17). The "*imaging of a fall*" or the "motion of falling bodies" is a principle of transformation that unites Milton's and Arnold's works and characterizes a possible analysis and interpretation (12). In effect, this transformation occurs through and by bodies in motion and contact with other bodies and environments around them.

Burke's imagery of transformation, especially as it involves bodies in motion, connection, and shared sensation, typifies the kind of *dynamis* that Aristotle's theories of rhetoric, art, and perception suggest. Like Aristotle's characterizations of rhetoric, art, and perception as all sharing the capacity of potential, transformation, and change, the imagery of transformation is crucial to rhetorical identification because it exemplifies the process of change. Burke realizes that his selection of texts exhibits "sinister implications of a preference for homicidal and suicidal terms" but argues that "the principles of development or transformation ('rebirth')" that these terms "stand for are not strictly of such a nature at all" (*A Rhetoric of Motives* xiii). The "imagery of slaying is a special case of transformation, and transformation involves the ideas and imagery of *identification*" (20).[10] In fact Burke aligns his terms for killing and rebirth with his overall aims for rhetoric in general. He elaborates, "We begin with an anecdote of killing, because invective, eristic, polemic, and logomachy are so pronounced an aspect of rhetoric," but he notes that "we can . . . always look beyond this order, to the principle of identification in general" (20). Identification does not deny the divisions or the wrangling of rhetoric but rather productively works to make connection. Identification is a tool of a wider scope in that it brings different motives, communicators, and bodies together in the aims of rhetoric.

As Burke suggests, Empedocles's body, especially in his transformative death, can be seen to function as "proof" of the worldview of interconnectedness among bodies, discourse, and matter that he espoused during his life. The story of Empedocles's death, as told by Heraclides, begins with Empedocles's disappearance after a great banquet. A servant reports hearing a voice calling Empedocles by name and a brilliant light appearing at the summit of Etna. Searchers climb Mount Etna looking for Empedocles but retrieve only one of his famous bronze sandals near the top of the mountain. Empedocles's disciple Pausanias calls off the search, "saying what had happened was a fulfillment of prayer; Empedocles had now become a god, and they should all worship him as such" (quoted in Lambridis 16). As Helle Lambridis explains, Empedocles, already well known for having performed several feats or miracles, desired to prove to his audiences his immortal status by making his body appear to have been lifted up by the gods. Various myths circulated in Greek culture of bodies lifted to heavenly realms—Iphigenia, Ganymede, and Oedipus all disappeared, never leaving a body behind. Lambridis assesses, "We can be fairly sure that Empedocles longed for a similar death; but living in a more rationalist age, he was not sure it would be granted to him. So he forestalled fate by creating his own legend" (18). Empedocles's body, or lack of body, once it is reported as becoming one with the elements, is potentially proof enough for his audiences of his worldview.

In addition Empedocles's own ailing body may have contributed to his desire to be remembered at the height of his powers. During his lifetime Empedocles was known for his legendary abilities: he helped overthrow an oligarchy, saved a town from a plague, changed the direction of a river, and purportedly brought a woman back to life (Lambridis 12). Lambridis suggests that toward the end of his life Empedocles may have been conscious of the fact that "he had given all of which he was capable" and that "his mental powers were on the wane" (18). From a disability studies perspective, Empedocles's suicide can be seen as an example of the ultimate in "overcoming" narratives; rather than face that his mental abilities were waning and his body was aging, Empedocles chooses to end his life to preserve the memory and legend of his abilities. Empedocles's suicide, however, is also a choice that reaches toward immortality while demonstrating and possibly proving the theories he had developed over the course of his life.

As Burke suggests, Empedocles becomes one with the elements and forces in his cosmology when he falls into the volcano. Empedocles's body, then, becomes bodies, as it intermingles with the various elements and forces through the theory of the mixture of all things via pores and effluences. Empedocles, by throwing himself into the volcano, proves his lifelong theories in the most convincing way possible, leaving no evidence to the contrary that all bodies are physically and tangibly interconnected entities. He returns his body to the elements and the forces that comprise all bodies and matter. In Empedocles's worldview, his body and the legends told and written about it prove that he has always been and will always be connected to other bodies via the theory of pores and effluences. In terms of Burke's rhetorical identification, the ultimate transformation takes place with Empedocles's act, as he becomes identified with his theories, his teachings, his words, and his world. Empedocles's self-immolation, mythical or not, and Burke's recounting of it in his formation of rhetorical identification demonstrate that multiple and diversely able bodies are integral to rhetorical identification. Burke's associating of Empedocles's body with the bodies of Milton and Samson and their corresponding blindnesses—their common sensations and their shared processes of consubstantiality—shows that both traditionally able bodies and disabled bodies comprise the bodies of identification.

Tactile Bodies: Proof, Persuasion, and Belief

As Burke describes, rhetorical identification offers an alternative to the "holds and the counter-holds, the blows and the ways of blocking them . . . for every proof [and] the disproof" of Aristotle's rhetoric (*A Rhetoric of Motives* 52). The bodies integral to identification in Burkean theory, in this way, prompt a revision of the kinds of "proof" and "disproof" that operate in traditional rhetorical appeals. By using Empedocles's body as an example, Burke suggests the possibility that bodies, movements, and common sensations can function as "proof" and "disproof" in rhetorical situations. If people are to believe what Empedocles believed in—that all bodies are made of the elements that intermingle and interanimate through pores and effluences—Empedocles must transform his body into the elements. Belief is dependent on what Burke would call a principle of transformation from Empedocles's one body to multiple bodies—belief transforms his body into multiple bodies via proof or disproof, depending on the situations and contexts of Empedocles's followers, his contemporaries, and even those learning about his legacy. In short, Empedocles's body ideally becomes belief itself, operating as proof.

In contemporary rhetorical theory, Empedocles's sense of connectedness among all bodies, beings, and words resonates with revisions of rhetorical situations that focus on enactments, affects, ecologies, structures of feeling, and the physical registers of rhetoric. Empedocles's teachings, including those lessons taught in his life and death, when considered in relation to Burke's naming of the "common sensations" involved in consubstantiality, demonstrate that touch is integral to rhetorical identification. By forwarding a rhetoric based on touch, it is possible to extend the efforts of recent refigurations of the rhetorical situation into a more total transformation for the needs and abilities of all rhetors who form rhetoric in physical and tactile relation with other bodies.

The various "proofs" of the body and bodies that Empedocles's example raises also inspire broader interrogations of the implications of the proofs in Aristotelian and contemporary rhetoric. As Kathleen Welch argues in *The Contemporary Reception of Classical Rhetoric,* Aristotle's three artistic proofs demand new translations for new readers. Exploring various translations of the appeals, especially "artificial," "artistic," and "proof," Welch asks, "How clearly do these words speak to contemporary readers of English?" (13). Noting that "proof" poses "difficulties in the post-empirical universe of discourse," she suggests "translat[ing] *pistis* as 'persuasion,' given the meaning 'proof' has acquired in the last two hundred years of empirical use" (13). Welch also suggests "interior" for "artistic," arguing that "the way would then be set for interpreting *ethos, pathos,* and *logos* with more of the psychological complexity . . . that occurs in everyday discourse" (13). Like Kennedy, she critiques the translation of *pisteis* as "proof" because it "bypasses the crucial Greek concept of belief that is present in *pisteis*" (139).[11] Although Welch favors the definition of "interior persuaders" for the appeals, it is possible to understand persuasion as functioning both internally and externally in relation to elements of Aristotle's theories of rhetoric that share crucial connections to the experience of sensation and perception, which is always an internal and external experience, and in light of Burke's theories of rhetorical identification, with its various pressures and counterpressures. In calling attention to how belief functions in the realm of proof, persuasion, and identification, redefinitions of the *pisteis* and reconsiderations of the rhetorical situation have the potential to respond more flexibly to a wider range of rhetors, especially those who use touch rhetorically.

From a disability studies perspective, reformulations of the *pisteis* are particularly necessary in order to value the rhetorical productions of disabled rhetors, especially those who use touch rhetorically, such as users of FC. As Brenda Jo Brueggemann and James Fredal note, existing beliefs about the "proof" or "persuasion" involved in representations of disability currently in circulation often limit the possibilities of disabled rhetors. Asking questions about *ethos*, *pathos*, and *logos* in relation to disability, they ask, "What of the character of the disabled person—or of those who persuade about disability, whether disabled or not?" (254). They wonder, "How is disability emotionally presented? (Why are its emotional tropes so limited to pity, depravation, degradation, inspiration and the like? What weight do these emotions carry in our culture?)" and "how . . . is a construction of disability established as logical?" (254).

Exploring these questions in relation to Keller's rhetorical productions reveals the ways in which certain beliefs about disability affect the use of the *pisteis*. Keller's various rhetorical situations, for example, were driven by limiting beliefs and stereotypical assumptions regarding her authenticity and credibility as a rhetor. She is doubted in part because of her disabilities and in part because no models exist for understanding her method of rhetorical production. Keller regularly experienced doubt from audiences regarding her *ethos*, especially her character and credibility. Aristotle states that a speaker's good character makes him credible and persuasive because "we believe fair-minded people to a greater extent and more quickly [than we do others] on all subjects in general" (*On Rhetoric* 1356a4; trans. Kennedy 38). Accused at least twice of plagiarism, Keller was not perceived as entirely credible by her audiences, who struggled to believe not only Keller but also the collaboration between Keller and Sullivan. Macy, Sullivan's husband and Keller's editor, observed that half of the people believed that "Annie Sullivan is just a governess and interpreter, riding to fame on Helen's genius," while the other half believed that Helen was "only Annie's puppet, speaking and writing lines that are fed to her by Annie's genius" (quoted in Lash 319–20). This split in belief among her audience, hinging on who believed whose ability, illustrates the assumptions regarding how touch functions in rhetoric. Simply put, Keller was not believed because she used touch to communicate, disrupting the familiar boundaries between the bodies involved in the roles of speaker/*ethos*, audience/*pathos*, and message/*logos*.

Keller's Rhetoric of Touch

From the perspective of the triangular model of the rhetorical situation based on clear divisions between speaker/writer, audience/reader, and message, Keller's rhetoric does not fall neatly into any one category. Her texts, multiple and multiply authored, often involve the physical presence of an audience in the process of rhetorical production. Keller's uses of the appeals of *ethos, pathos,* and *logos* are similarly blurred. Rather than *ethos* residing in one body of the rhetor or *pathos* being associated primarily with the emotions of an audience, Keller's *pisteis* are multiple and mobile. Her *ethos* as a writer is not located in one node, separated from her audience or message. Nothing exemplifies this more than her relationship with Sullivan, whose being Keller reports as feeling "inseparable" from her own, an attachment that is built into the construction of her *ethos.* Her *logos*—her written messages and logical constructions—often consist of what she calls "borrowed feathers," from the hands and words of those with whom she has interactions, not wholly distinct from her own being. These interlocutors resemble an audience in some respects but in other ways shape the substance of her own *ethos* and messages as a writer. Because of these blurrings, Keller's message or *logos* was compromised throughout her writing career because her writing is not easily identifiable as the product of one able and independent rhetor. Her *ethos* is doubted, and the emotional connections she may have with an audience are consequently questioned. The traditional shape of the rhetorical situation and the bounded nature of the appeals as they are typically conceived do not fit Keller's rhetorical production.

From a perspective of rhetorical production attentive to sensation, however, Keller's tactile rhetoric generates effective messages, forms meaningful identifications, and facilitates opportunities for rhetorical production across bodies. Touch is integral to Keller's use of the appeals of *ethos, pathos,* and *logos.* Rhetorical production for Keller took place in the generative space of touch, existing at first between two bodies and then a wider range of bodies and entities as her audience widened and as writing technologies developed.[12] A rhetoric attentive to touch also more adequately captures Sullivan's contribution to the rhetorical productions she undertook with Keller. According to Nielsen, Sullivan "leaned on Keller," and Sullivan's "lifelong struggle with chronic illness and depression was far more debilitating than Keller's deaf-blindness" (x). From this perspective, which resists understanding Sullivan as either able-bodied assistant or puppeteer, Sullivan and Keller's *ethos* is coconstructed.

Both conceptually and substantially, Keller and Sullivan share an *ethos* and are identified with each other. Keller's *ethos* especially transgresses the accepted understanding of the rhetor as the individual autonomous person, located in a singular, hard-edged body. Keller reports that she gained a sense of herself as someone with a sense of *ethos,* with cognition and emotion, only in relation to and in touch with other people and things. She writes, "Before my teacher came to me, I did not know that I am. I lived in a world that was no-world. I cannot hope to describe adequately that unconscious, yet conscious time of nothingness . . . I had neither will nor intellect . . . I can remember all this, not because I knew that it was so, but because I have tactual memory. It enables me to remember that I never contracted my forehead in the act of thinking. I never viewed anything beforehand or chose it. I also recall tactually that never in a start of the body or a heart-beat did I feel that I loved or cared for anything. My inner life, then, was a blank without past, present or future, without hope or anticipation, without wonder or joy or faith" (*The World I Live In* 113–14).

In her reflections of her life, Keller often implicitly employs a Burkean concept of "substance" and its absence to mark the differences between her experiences before and those after she began to use touch rhetorically. Recalling her past and referring to herself in the third person, she writes, "She did not know 'shadow' because she had no idea of 'substance'" (*Teacher* 42). Before she met Sullivan, Keller could not conceive of herself as a self. Choosing words that lack substance, she describes herself as having a "blank" inner life. Without language and without meaningful connection to others, she initially lacked the ability to cultivate character, express her feelings to others, exert her intellect, register emotions, or make sense of her world. In fact her sense of self, her emotions, and her cognition did not "register" until she formed meaningful connection, through touch, with Sullivan at the well by her house and her "tactual memory" catalyzed kinship with the rest of the world.

Once Keller discovered her world and herself through touch, she began to use language rhetorically, forming identifications with others through consubstantiality based on touch. In these connections with others, she developed her appeals of *ethos* and *pathos,* which, rooted in touch, are never singular. Keller conceived of her entire being and body through touch, writing, "Sometimes it seems as if the very substance of my flesh were so many eyes looking out at will upon a world new created everyday" (*The World I Live In* 29). In the Burkean sense of substance, Keller was connected through and to her world, internally and externally. Merging visual and tactile senses, she describes the world of the blind as a "tangible white darkness" (*The Story of My Life* 35). To make her way through this tangible darkness, she formed connections and identifications. Keller was substantially connected to her audience; touch shaped her notion of audience as much as it shaped her notion of herself and her *ethos:* "I am always intensely conscious of my audience. Before I say a word I feel its breath as it comes in little pulsations to my face" (*Midstream* 214). Touch guided each element of Keller's rhetoric, a process that enabled her to form meaningful connections and identifications with her readers and other interlocutors. Keller's rhetoric was a rhetoric of touch—a physical and psychological system of expression that enabled her to form emotional, logical, and personal connections and potential identifications with others.

Keller's rhetoric of touch particularly shaped her development as a writer, although Keller initially struggled to understand her composition process. Describing her writing as a "not yet complete process," she relates, "It is certain that I cannot always distinguish my own thoughts from those I read, because what I read becomes the very substance and texture of my mind. Consequently, in nearly all that I write, I produce something which very much resembles the crazy patchwork I used to make when I first learned to sew. . . . Trying to write is very much like trying to put a Chinese puzzle together. We have a pattern in mind which we wish to work out in words; but the words will not fit the spaces, or, if they do, they will not match the design. But we keep on trying because we know that others have succeeded, and we are not willing to acknowledge defeat" (*The Story of My Life* 67–68). Keller found the writing process to be a pattern that did not fit the space she imagined. Her sense of logic or *logos* was tactile—what she read became the "very substance and texture" of her mind. Like Burke's sense of substance, Keller's substance was both internal and external. She felt pressure to control her words, as other people do, yet acknowledges that this was a difficult task, as she attempted to "marshal the legion of words which come thronging through every byway of the mind" (67). The sense of touch guided her approach to *logos* as much as it did her *ethos* and *pathos;* the pathways of Keller's mind were shaped by her physical interactions and identifications with others.

In *Midstream,* written midlife, Keller seems better positioned to accept that her process as a writer may be different from others and that this process can be generative: "Perhaps it is true of everyone, but it seems to me that in a special way what I have read becomes a part of me. What I am conscious of borrowing from my author friends I put in quotation marks, but I do not know how to indicate the wandering seeds that drop unperceived into my soul. . . . I prefer to put quotation marks at the beginning and the end of my book and leave it to those who have contributed to its interest or charm or beauty to take what is theirs and accept my gratitude for the help they have been to me" (328). Touch shaped every aspect of Keller's rhetorical production and every element of her rhetorical situation. Particularly generative for Keller was her approach to the "ownership" of her writing. Keller shared her writing with everyone with whom she had come in contact; proof is inconsequential. Touch shaped her rhetoric and secured her sense of identification with her collaborators and audiences. In Burkean terms, she was fully identified, in sensation and texture, attitude and belief, content and form, with the collaborators and audiences she had formed and by whom she had been formed. From the perspective of Empedocles's contributions to rhetoric, Keller embodied a living example of the theory of pores and effluences, using touch to intermingle ideas, messages, feelings, and bodies. In Aristotelian terms, she was exercising a practice of rhetoric as a *dynamis*, exploring the capacity of rhetoric to grasp connections with others in discursive and nondiscursive ways.

Rhetoric Reshaped

A theory of rhetoric that attends to touch is necessary to understand and value the rhetorical productions of rhetors such as Keller, users of FC, and other people with disabilities who interact physically with others, disabled or not, in order to create rhetoric. Current theories and modifications of the rhetorical situation and models of collaborative rhetorics, although productive for a wider range of rhetors and contexts, do not fully account for the needs and abilities of rhetors with disabilities who compose in tactile relation to other bodies. As Sharon Crowley explains, modern rhetorical theory and composition studies assume "that 'actors' 'send' 'messages' to 'receivers' through 'channels,'" ignoring the critical fact that "discursive transactions are embodied . . . because both fields still cling to liberal-humanist models of the speaking subject—a sovereign, controlling disembodied and individual voice that deploys language in order to effect some predetermined change in an audience" ("Body Studies" 177). For rhetors with disabilities and their collaborators, this assumption does a particular disservice.

In addition, as Brueggemann and Fredal suggest, stereotypes about disability and the disability experience often affect how rhetors with disabilities use the appeals associated with this model of communication, especially how people with disabilities can develop an *ethos* or a sense of character and credibility, create appeals to *pathos* via emotional connections with audiences, and mobilize messages or logical constructions about disability. A triangularized model of communication is particularly inaccessible for rhetors on the autism spectrum who are viewed from a medical model perspective. As Paul Heilker and Melanie Yergeau explain, the "autist, as medically constructed, is self-focused, a two-pointed rhetorical triangle floating outside the context bubble" ("Autism and Rhetoric" 494). As Keller's experience and the experiences of those who use FC show, the *ethos* of rhetors with disabilities are often doubted when they produce rhetorics based on touch and outside of the mold of the independent body, a skepticism that affects how they may use appeals of *pathos* and *logos*.

To posit touch as rhetorical—a theorization that supports the alternative rhetorics of Keller and the additional disabled rhetors whose communications populate the following chapters—I offer a revision of the rhetorical situation that moves away from the preference for an independent, able rhetor and audience using primarily verbal and discursive means, and toward more dynamic interplays of rhetorical identifications that imagine multiple agents of diverse ranges of discursive and nondiscursive abilities who may or may not be completely independent and autonomous. This theory of rhetoric contributes to ongoing efforts focused on interrogating the assumption of the individual rhetor and revising the presumption that author, message, and audience are separate and independent entities. As Jack Selzer, exploring possibilities for rhetorical bodies, asks, "How might the articulation of a material rhetoric force us to reconceive rhetorical entities like 'speaker,' 'writer,' 'arrangement,' and 'audience?'" (10). Rhetorical touch contributes to an articulation of the material force of rhetoric and restructures the assumed separateness of certain rhetorical entities. The appeals, considered especially in relation to the sense of touch, reveal rhetorical opportunities for identification and connection across contexts of disability, exemplifying the possibilities of Aristotle's *dynamis* of rhetoric.

Specifically, I propose bringing the traditional model of rhetoric as the triad of three elements—speaker/writer/*ethos,* audience/listener/*pathos,* and message/text/*logos*—in contact with three sophistic elements of rhetoric that I argue in the remainder of this book are tactile arts: Empedocles's felt *logos* and the practices of *mētis* and *kairos.* Taken together, these elements form two rhetorical triangles always in contact: Empedocles's felt *logos, mētic-ethos,* and *kairotic-pathos* intermingle and interanimate with the rhetorical triangle of speaker/author, audience, and message and with the Aristotelian triad of *ethos, pathos,* and *logos,* keeping the elements of rhetoric in dynamic interaction and poised for new potential.

With the formation of two triangles always in contact, this representation of the theory of rhetoric reshaped encourages multiple points of contact among bodies and rhetoric. Rather than the elements of a rhetorical situation occupying a singular point or angle of a triangle, the placement of the triangles along their sides or at their shared points of contact is designed to encourage ways of understanding the situations and elements of rhetoric as connected, interdependent, and contingent on each other. By bringing the two triads together, the shape of a rhombus or diamond emerges. In this way the figure can be understood as two triangles coming together, whether at their sides or points, as a whole. This figuration recalls the most basic properties of touch: an area that is both joined and separate, a flexible and contingent whole that is greater than the sum of its parts.

In the figure in the middle, for example, the sides of the triangles featuring *ethos* and *mētis* are aligned with each other, in contact along their sides. This configuration is designed to accommodate not only multiple rhetors but also a spectrum of rhetors of different abilities. This configuration supports the efforts of rhetors such as Keller and others who work in close physical contact with others to produce rhetoric, forming an *ethos* that is multiple and interdependent. The other elements of the triangles are also in contact: *logos* and Empedocles's felt *logos* come together at their angles but also share the entire left side of the figure. Similarly *kairos* and

The Rhetorical Situation/ Appeals

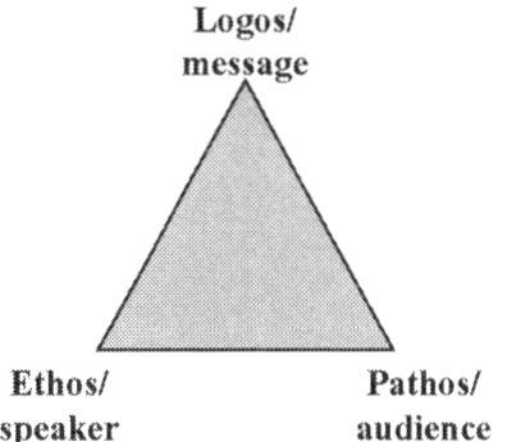

Rhetoric Reshaped for a Rhetoric of Touch

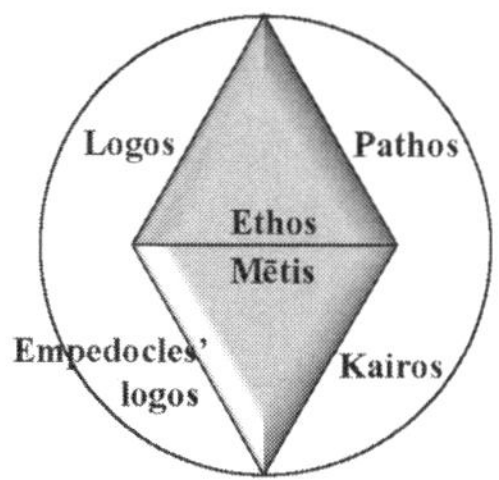

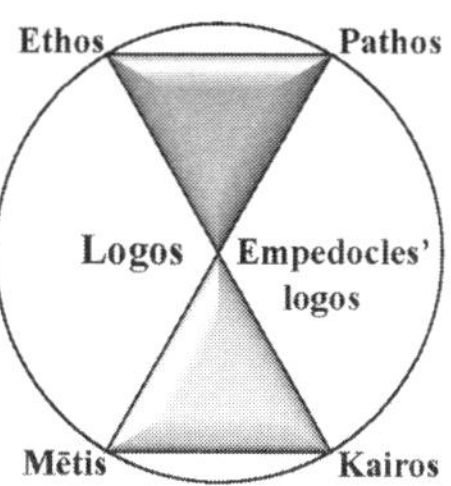

A rhetoric of touch. The figure on the left represents the traditional rhetorical situation and appeals. The figures on the right illustrate the possibilities of reshaping rhetoric to support rhetorical touch.

pathos join primarily at their angles but also share the entire right side of the figure when the figure is thought of as a whole. "Rhombus" comes from the Greek meaning something that is "spun" and "to turn round and round" (Liddell and Scott). From this perspective the figure invites different positioning and dimensionalities that may emphasize certain elements of a situation or identification. Spinning the figure one way or the other, for example, produces a figure in which, as an alternative to *ethos* and *mētis* sharing a middle, *logos* and Empedocles's *logos* occupy the center space. Or, through a different turn of the figure, *pathos* and *kairos* are situated in the central location. These alternative configurations can accommodate a wide range of contexts and situations, forming a figure in which the shared elements of rhetoric flexibly interchange places. In this sense, in the event that a situation or identification necessitated specific focus, even a configuration such as the figure on the right is possible, in which *logos* and Empedocles's *logos* meet in the middle at a center point, with the other paired elements occupying the sides of the figure. Similarly a figure is possible in which *pathos* and *kairos* or *ethos* and *mētis* share a center point, with the other elements forming the edges. In these cases, however, it is important to be mindful of the fact that even the point at which these elements come together is not singular but is always multiple in its contact.

Surrounding the figures at all times is a circle that denotes the contexts and contingencies of embodiment that suffuse the elements of any given situation, interaction, communication, and identification. The circle derives from the connection that Aristotle makes between two words in his *Rhetoric* and the implications of this connection in the context of disability. Explaining the differences between probable and irrefutable signs and statements, he invokes the term *tekmērion,* a kind of sign that people offer when thinking that a "matter were shown and concluded" (*Rhetoric* 1357b17; trans. Roberts 43). As Aristotle notes, "*Tekmar* and *peras* ["limit, conclusion"] have the same meaning in the ancient form of [our] language." Marcel Detienne and Jean-Pierre Vernant, theorizing this remark by Aristotle, offer a connection between the limits and possibilities of language operating as sign, indication, and guidepost (288). The "sign" meaning of *tékmar* or *tékmōr,* for example, is part of "the same terminology" as "*peîrar,*" "*péras,*" and "*peráō,*" meaning "path," "to cross," and can refer to a route of passage or the strategy that opens up a way through an uncertain space or journey (288–89). This connection opens up language and its possibilities rather than signaling a conclusion or settling of it. Furthermore "in the form of *péras* the same word is used in medical terminology to refer to the end of a bandage or piece of material which surrounds a wound or protects a limb" (290). For Detienne and Vernant, this use of the bandage stands for the complexities, contradictions, and possibilities of the circle and the bond in Greek thought and culture: "Here we find one particular type of path which takes the form of a bond which fetters and, conversely, the action of binding is sometimes presented as a crossing, a way forward" (291). These elements of embodiment and language are represented by the circle enclosing the figures of rhetoric reshaped, signaling that disability can be a way forward even as it is perceived as a limit in culture and society and that communication and bodies are always coconstitutively marked.

As the following chapters bear out, it is possible to understand each of the elements of rhetoric reshaped as tactile and as the product of bodies in contact. Touch is integral to each of the three sophistic terms and practices I pair with the traditional elements of the rhetorical situation. Empedocles's notion of a "felt" *logos,* for example, is consistent with his overall theory of the cosmos and general worldview of beings, matter, and discourse as tactile. *Logos,* for Empedocles, also participates in the physical process of intermixing and intermingling via his theory of pores and effluences. Words, especially in repetition, have the power physically to touch and move rhetors and audiences in Empedocles's understanding, resulting in a redefinition of *logos* beyond traditional notions of rationality. Traditional *logos,* with its connotations of logical reasoning, is an appeal that is often positioned as out of reach for rhetors with psychological disabilities and differences. By exploring Empedocles's concept of *logos* as a felt experience, there exist possibilities for accommodating the needs and abilities of a wider range of rhetors.

Touch also operates crucially in the sophistic rhetorical practice of *mētis,* or cunning and embodied intelligence. This tactile *mētis,* especially when understood as a habit, can be valued as integral in shaping the formation of *ethos* in sophistic rhetoric and Greek culture and society, contributing to a redefinition of *ethos* beyond the confines of the singular, independent body. *Ethos* reconfigured this way is particularly supportive of rhetors with cognitive differences who form their character in physical connection with others or who use forms of mediated communication such as FC. Although traditional and contemporary understandings of *ethos* often do not support the needs or abilities of rhetors with autism, for example, redefining *ethos* in relation to the tactile practice of *mētis* affords a conceptualization of *ethos* that is multiple, neurodiverse, and physical.

Finally, a tactile sense of *kairos* based on the proximity of bodies offers potentialities for understanding *pathos* as an appeal shared among bodies in close contact. The tactile and technical properties of *kairos* resonate in sophistic notions of *kairos* as a dynamic rhetorical situation consisting of diverse types of bodies in motion and model alternative emotional interactions among these bodies. Appeals of *pathos* based on a *kairos* of proximity enable new possibilities for understanding disability outside of normative emotional paradigms of pity, inspiration, and sorrow. Rhetors with physical disabilities are especially susceptible to encountering audiences who expect emotional appeals based on pity and inspiration because these emotions dominate messages about the disability experience in tropes of personal tragedy and overcoming narratives. Exploring emotion in relation to proximity, touch, and interconnectedness reveals a range of diverse emotional possibilities for expressing the disability experience.

By bringing the Aristotelian appeals into contact with Empedoclean and sophistic rhetorics, especially through the rhetorical practices of *mētis* and *kairos,* I draw on possibilities and commonalities typically neglected in the rhetorical tradition. As Detienne and Vernant explain, some elements of Aristotle's philosophy are not entirely incompatible with the "indirect and groping knowledge" of the sophists (313). Specifically, there "are good reasons for believing that the theory of prudence expounded in the *Nicomachean Ethics* expresses a desire to embrace . . . the traditions of the orators and sophists and the types of knowledge which are subject to contingency and directed towards beings affected by change" (316). In this interpretation, Aristotle's understanding of *phronesis* incorporates elements of transformation and change not unlike those in his definitions of "art" and "touch." Contingent, flexible, and responsive knowledge is prized in this formulation.

In addition, as Richard Leo Enos has pointed out in "Aristotle, Empedocles, and the Notion of Rhetoric," certain elements of the sophistic tradition, such as its "indirect heuristic processes and procedures[,] advanced knowledge through expression in ways which, while not in harmony with Aristotle's views of rhetoric, must be taken as a non-rational epistemology that made contributions to knowledge which both preceded and survived Aristotle's notion of the 'art' of rhetoric" (20–21). As a rhetoric rooted in nonrationality, touch is an indirect knowledge system that bridges different ways of knowing. This approach to knowledge creates potential for connections between Empedoclean, sophistic, and Aristotelian rhetorics. As Burke's evocation of both Aristotle and Empedocles in his theory of rhetorical identification shows, these potential connections in ancient rhetorics are generative.[13]

Taken together, the dynamic tactile elements of rhetoric in Empedoclean, sophistic, Aristotelian, and Burkean rhetorics modify elements of the rhetorical situation for the needs of rhetors who rely on touch. Each of the examples of rhetorical production through touch in the following chapters exemplifies the potential of rhetorical identification—partial and connective—specifically between people with and people without disabilities and between people of varying and diverse levels of disability. In the spirit of Burkean identification and Aristotelian *dynamis,* people with disabilities use rhetorics based on touch as a process of change, transformation, and potential. A rhetoric of touch, for example, supports understanding of a rhetorical partnership such as Keller and Sullivan's, based on belief instead of doubt. A rhetoric rooted in valuing touch positions Keller as a credible and trustworthy rhetor rather than one mimicking her teacher's intelligence or hoodwinking audiences. A model of a tactile rhetoric based on the connections between and among bodies supports the transformative feelings and emotions that can circulate among rhetors and their audiences. Rhetors who use mediated communication devices such as FC and their facilitators are best served by a model of rhetoric that values touch as transformative rather than as suspicious. In short, a rhetoric of touch reshapes the appeals of *logos, ethos,* and *pathos* for a wider diversity of rhetors and creates possibilities for identification among a range of contributors who possess a spectrum of abilities and disabilities.

In chapter 2, I explore identification in more detail as a site of touch. Delving into three sites of touch in phenomenological and deconstructionist philosophy—the *chiasmus* (an intertwining), the limit, and the fold—I position touch as a location for connection across material-discursive and social-individual registers. These sites of touch offer methods for transforming rhetorical situations into rhetorical identifications, suggesting ways to intertwine the elements of the rhetorical situation, to test the limits of the terms and the bodies to which they refer, and to fold new terms and positions into the triads. After locating sites of transformation and identification via touch, I fold, delimit, and intertwine the *pisteis* with the tactile practices of Empedocles's felt *logos, mētis,* and *kairos.*

2

Locating Touch

The Substances and Spaces of Rhetorical Identification

The lived experience of disability—particularly the wide range of difference that accompanies physical, cognitive, and psychological disability—often poses challenges to existing theories of identity and identification. It is often assumed that people with different kinds of disabilities may have trouble identifying with each other. People with a cognitive or mental disability, for example, have experiences, interests, and needs that often differ from those of people with physical disabilities. Differences such as these can impede identification and can result in horizontal hostilities among people of the same group or identity category. As Eli Clare explains, "Marginalized people from many communities create their own internal tensions and hostilities, and disabled people are no exception" (92).

People with disabilities may also have trouble initiating identifications with nondisabled or temporarily able-bodied people. Prevailing stereotypes about the disability experience, for example, which position disability as a personal tragedy, individual defect, or occasion for inspiration, limit the ways that nondisabled and disabled people can connect with each other. As Krista Ratcliffe has explored in the context of cross-cultural communication, strained identifications such as these are "troubled identifications," or identifications "troubled by history, uneven power dynamics, and ignorance" (47). In the context of disability, troubled identifications arise from unequal power relations between disabled and nondisabled people, a history in which disabled people have been ignored and discriminated against, and unexamined stereotypes about the disability experience.

Although these troubled identifications exist, there are contexts and moments in which identification succeeds. For example, in a key moment of disability rights history, the 1977 occupation and sit-in of a federal building in San Francisco, people with a wide range of different disabilities and identity categories joined together to achieve identification and enforce legislation. During the sit-in, over a hundred disabled people took over the U.S. Department of Health, Education and Welfare (HEW) building for approximately one month, demonstrating for the enforcement of Section 504 of the Rehabilitation Act of 1973. Smaller protests and occupations included those in Washington, D.C., where protesters demonstrated inside the

HEW building and the office of Joseph Califano, appointed by President Jimmy Carter as secretary of HEW.

The sit-in is typically considered a key victory for the disability rights movement. Alliances formed during the demonstration fostered a coherent sense of identity among people with disabilities. For the first time, people with a wide range of disabilities—physical, sensory, cognitive—came together in a shared cause, forging a new group identity to achieve significant results. As Ed Roberts, a leader of the disability rights movement, related, "To see hundreds of people with disabilities rolling, signing, using canes, the more severely retarded people for the first time joining us in this incredible struggle, is one that leads me to believe that we're going to win this" (*The Power of 504*). From a legal perspective, passage of 504 and the enforcing of its regulations meant that any entity receiving federal funding—schools, universities, libraries, government offices—must be accessible to people with disabilities. From a rhetorical perspective, the sit-in also represented a victory of identification, as many people with diverse disabilities identified their interests with each other and even with a larger nondisabled public.

Although heralded as passage of a bill of rights for disabled people, the victory was hard won. It is often told as a "success story" of the movement, but the demonstration was one born out of years of frustration and rhetorical failures. As Sharon Barnartt and Richard Scotch describe, in the years leading up to the demonstration, deliberation had stagnated; three years of letter writing and lobbying had produced no results (164). The Carter administration had done virtually nothing in four years to enforce the first major law that banned discrimination against people with disabilities. Even once demonstrators took over the HEW building, over a month passed before any legislative action.

Demonstrators felt forced to implement a wide range of tactics, some overtly antagonistic, in order to get their message across. In rhetorical terms, protesters moved from conciliatory rhetoric of the open hand to a more muscular rhetoric of the closed fist (Corbett.) As one protester said, describing the decision to take over the federal building, "It's the first really militant thing disabled people have ever done. And we feel like we're building a real social movement. We want people to listen to us. We have tried negotiations. They do not work" (*The Power of 504*). Listening, however, was not simply a matter of having one's voice heard; demonstrators wanted people to *experience* their arguments. Protesters turned to their bodies to convey their message and to identify with their audiences through experience. After they took over the office of an HEW regional manager, for example, blocking his access to the bathroom, one protester described the event in coercive terms of identification: "We have had to learn all of our lives to control our bladders. You must learn that lesson now, too" (*The Power of 504*). These strategies of identification were designed by demonstrators to make audiences feel, on a bodily level, their frustration and the experience of what it was like to be disabled in America.

Protesters, however, also rebuked initial attempts by their audience to assume they understood what it felt like to be disabled. When HEW secretary Joseph Califano attempted to respond to the protesters, he began by saying, "I know how you feel" but was immediately interrupted by protesters, some saying and some signing, "No, you don't!" and some simply shaking their heads and gesturing to their wheelchairs ("504 Demonstrations"). With these actions, demonstrators drew attention to the fact that many were risking bodily injury by staying in the building for as long as they did, forgoing necessary assistance and support by trained aides and attendants (Shapiro 67). One woman with a spinal cord injury reported that she needed to be turned over in her bed once a night; she had to rely on other demonstrators, untrained in how to do this, to turn her (*The Power of 504*). Protesters slept on floors, lacked hot water, and waited two weeks before showers were outfitted for their needs. The demonstration was difficult physically and psychologically; protesters wanted their audience to feel this hardship with them, rather than assume they understood.

Demonstrators eventually secured the chance to speak to an audience of legislators in specially convened congressional hearings, but the gulf between them and their audience only widened. The activist and future assistant secretary of education Judy Heumann attempted to describe discrimination against disabled people, calling it "so intolerable that I can't quite put it into words" (*The Power of 504*). She continued, "I can tell you that every time you raise issues of separate but equal, the outrage of disabled individuals across this country is going to continue; it is going to be ignited. There will be more takeovers of buildings until finally maybe you begin to understand our position. We will no longer allow the government to oppress disabled individuals. We want the law enforced. We want no more segregation. We will accept no more discussion of segregation" (*The Power of 504*). In an interaction that echoed the encounter of 504 demonstrators with Califano, Heumann admonished her audience for making assumptions about identification. When nondisabled members of the congressional audience nodded their heads, trying to empathize with her words, Heumann struggled through her tears, stating firmly, "And I would appreciate it if you would stop shaking your head in agreement when I don't think you understand what we are talking about" (*The Power of 504*). In response, a group of disabled people clapped behind her.

What Heumann was trying to put into words is the feel of the disability experience. She was trying to communicate what it feels like to be excluded from participation in society and to be frustrated in attempts to convey this experience to a nondisabled public and legislature. While her testimony solidified broader identifications among different people with disabilities—she transitioned from speaking about "disabled individuals" to the repetitive use of "we"—it also spoke to the gulf of identification that separated this group from nondisabled people. Although the demonstration for 504 was successful, the rhetorical exchanges in its aftermath suggest that even as disabled people, including people of a wide range of different disabilities, came to identify with each other, they still experienced difficulty identifying with larger, nondisabled audiences.

Even among disabled people, the achievement of identification seemed fragile. Despite the productive group stance that people with disabilities cultivated during the 504 sit-in, there were signs that the model of identification they forged may not endure. As the occupation concluded, one activist, Ron Washington, conveyed reluctance to leave the building, explaining, "There's some hesitancy because of the relationship that was developed here—the comradeship around political needs and working together to get those needs taken care of" ("Handicapped Win Demands"). The frustration that activists such as Heumann expressed when testifying in congressional hearings also points to the challenges regarding identification across wider audiences, including nondisabled audiences. As Barnartt and Scotch describe, the 504 victory was largely a "symbolic" success with "real importance to the entire disability community," but "empirically measurable changes" resulting from the victory are difficult to locate because larger infrastructure, resources, and support necessary for lasting change were not yet in place (167–69). To a certain extent, the problems that activists such as Washington and Heumann articulated in the aftermath of 504 continue as obstacles that inhibit identification across wider ranges of audiences, especially nondisabled ones.

Questions of disability and disability identity have rarely been put in terms of the rhetorical concept of identification; yet identification offers a productive angle on complicated issues of identity and embodiment in disability studies and the perception of disability in wider culture. Rather than identity, which can be interpreted as focusing on a singular body or a single aspect of a person's or group's identity, identification emphasizes the multiple, flexible aspects and locations through which a person or group can identify with someone or something.[1] For Kenneth Burke, one way of creating identification among people and their interests is through consubstantiality—literally and symbolically the sharing of a common "substance"—a process that I position throughout this chapter as a sensory location that invites rhetorical identification through touch. Unanchored to a single body, identification through tactile consubstantiality offers flexible, malleable, and diverse engagements with the body and other bodies—disabled, nondisabled, or temporarily able-bodied. This tactile approach to consubstantiality addresses the complexities of identity and identification among differently disabled and nondisabled people such as those exhibited in the 504 demonstration and its aftermath.

Attention to the sensory elements of identification also extends existing theories of identification in rhetorical studies based on binaries and metaphors that are unproductive for people with disabilities and the wide audiences with whom they seek to establish identification. Specifically, I extend Burkean theories of identification and substance with meditations on touch offered by the philosophers Maurice Merleau-Ponty, Jean-Luc Nancy, and Gilles Deleuze, who engage with questions of bodily vulnerability and who locate touch as a material and discursive process of intertwining, delimiting, and folding among individual bodies and larger social groups. I then reread the 504 demonstrations via this rhetoric of touch, turning to the efforts of protesters such as Joan Tollifson, who uses the lessons of identification she learned during the sit-in as a model for interacting with other people, folding the sense of touch into her physical and rhetorical encounters in ways that blur but do not necessarily dissolve the boundaries between rhetor and audience and between disabled and nondisabled. I also explore how people commemorate the 504 protest online, folding rhetorical identification through consubstantiality into their current daily lives. In this way I rhetoricize the sense of touch in the experience of disability, locating its function as a material and discursive potentiality for new identifications and consubstantialities among people with differing disabilities, nondisabled people, and the temporarily able-bodied.

The Difficulties of Disability Identification

Over thirty years after the 504 demonstration, there is increased public awareness of disability issues, interdisciplinary academic programs of disability studies in universities, and significant additional legislation in the movement for disability rights. As Joseph Shapiro notes, citing advances in technologies, new civil rights protections, and an emerging political consciousness, "Never has the world of disabled people changed so fast" (4). Gulfs of identification still exist, however, among people with and without disabilities as well as among people with different disabilities. As Simi Linton notes, "Theories are needed . . . that conceptualize disabled and nondisabled people as complementary parts of a whole integrated universe" (*Claiming Disability* 120).[2] Shapiro states simply that "nondisabled Americans do not understand disabled ones" (3). Mary Johnson of the *Disability Rag* remarks, "Our society still does not understand disability rights. And it shies away from the experience as much as ever, if not more. . . . We in the disability rights movement still have a very large job ahead of us" (viii–ix). A clear priority of this job is increasing opportunities for identification among the disabled, nondisabled, and temporarily able-bodied. New opportunities for identification are needed not only between people with disabilities and the nondisabled but also among various people with different disabilities.

In disability studies, theorists and scholars have tended to focus on identity, rather than the rhetorical process of identification, to explore opportunities for connection. Lennard Davis, exploring the possibility that disability is the identity category that can unite all other identities, asks, "What is more representative of the human condition than the body and its vicissitudes? . . . Is there not something to be gained by all people from exploring the ways that the body in its variations is metaphorized, disbursed, promulgated, commodified, cathected, and de-cathected, normalized, abnormalized, formed and deformed?" ("Introduction" xv–xvi). Disability identity, however, may be particularly difficult to claim or to build identifications on because the experience of disability is so widely varying, indeterminate, and even contradictory. As Michael Bérubé describes, disability's "multifarious and indeterminate meaning" identifies "thousands of human conditions and varieties of impairment, from the slight to the severe, from imperceptible physical incapacity to inexplicable developmental delay. It is a category whose constituency is contingency itself" (vii–viii). Even among people with the same condition, disability is widely variable: the "question of 'who' qualifies as disabled is as answerable or as confounding as questions about any identity status" (Linton *Claiming Disability* 12).

Furthermore, as Davis describes, the firm sense of identity that first united people with disabilities as a distinct group, in which "identity was tied to the body, written on the body," has been replaced with more complex questions (*Bending Over Backwards* 13). In the decades following demonstrations such as the one for 504, a "second wave" of disability studies involves "questioning already occurring areas of identity formation" and "the differences (rather than the similarities) between impairments" (Davis "Preface" xiii). This questioning necessarily involves retheorizing the position of the body in disability identity. Although disability may no longer be tied exclusively to the body, the body is still important to theorizing disability. As Tobin Siebers argues, "Identities, narratives and experiences based on disability have the status of theory because they represent locations and forms of embodiment from which the dominant ideologies of society become visible and open to criticism" (*Disability Theory* 14). The body remains important to identity in disability, even as the category of identity has transformed.[3]

In rhetorical studies, existing models of identification do not accommodate the identities of people with disabilities or identifications made possible by the lived experience of disability. People with disabilities such as those who protested for 504 live their lives and create discourse often in direct physical connection and collaboration with other people, and yet existing models of identification often do not account for the material realities of this relational process. Although the protesters for 504 used identification to merge individual and collective interests on discursive and material levels, identification as it is typically understood relies inordinately in binaries and metaphors that separate these categories. As Ratcliffe has explored, identification often falls into an unproductive false binary emphasizing either individual or discursive determinism. Modern rhetorical theories of identification such as Burke's are thought of as strategies that "champion the idea of a human agent trying to control language for his own ends," imagining identification as a "space wherein the substance of one person is synthesized with substances of other people in order to bridge differences and create common ground for persuasion and political action" (52). Alternatively, postmodern theories of identification, often influenced by theories from Michel Foucault, Jacques Derrida, and Judith Butler, are typically understood as tending to diminish personal agency and focus on the "agencies of discourse and cultural structures in the construction of subjects," often questioning "the possibility of substance as well as the possibility of shared substance or common ground" (52). Identification is primarily a discursive process in this view.

This binary leaves little opportunity for identification in the context of disability. Identification is either substantial or discursive; the subject (and implicitly the subject's body) is either a completely able and autonomous entity or is fairly inconsequential in the process of meaning-making.[4] Identification is either individual or social: an autonomous, individual agent initiates identification or a person is identified through a larger social process via discourses and various cultural structures. Troubled identifications such as those between people with different disabilities or between nondisabled and disabled people are particularly difficult in this binary. As Ratcliffe describes, "Modern theories of identification often foreground personal agency and commonalities while backgrounding differences; postmodern theories of identification and disidentification often foreground differences while backgrounding personal agency and/or commonalities" (48). In the context of disability, the prizing of commonality based on personal agency or the championing of difference based on cultural or discursive structures is especially problematic.

People with disabilities, whether in the demonstration for 504 or in contemporary arguments for equal rights, maintain different values that do not rely on binaries of individual agency or social-discursive production. As Paul Longmore declares, disabled people prize "not self-sufficiency but self-determination, not independence but interdependence, not functional separateness but personal connection, not physical autonomy but human community" (222). Disability theorists such as Mairian Corker and Sally French, drawing from Peter Leonard's critique of postmodernism, yearn for models of disability discourse that do not rely on a binary that "effectively ask[s] us to choose between the extremes of 'reality determinism' and 'discourse determinism'" ("Reclaiming Discourse" 3). Resisting ways of thinking based on binaries or dichotomized oppositions, the disability community seeks to establish values that foster connection among its members and the nondisabled community.

As Ratcliffe describes, the binary between modern and postmodern theories of identification leads to a narrowing of potentials for commonality and difference among individuals that is usually represented in a visual metaphor or mental construct. Modern theories often "imagine identification as a metaphoric space," and visually "this process is often represented as interlocking circles with a shared middle space of common ground signifying a place of identification" (52). Postmodern theories situate identification as a discursive process "whereby subjects, because they have lost the object of their desire, metaphorically construct mental images of those objects and then act upon those images" (52). Although theorists in rhetoric and composition studies such as Jacqueline Jones Royster, Keith Gilyard, and James Berlin have interrupted the modern/postmodern binary and "mapped spaces for a rhetorical agent who possesses personal agency (albeit limited) even as that agent is socialized by enveloping cultural discourses," Ratcliffe notes that more widely the "reformist potential" of identification has been ignored (52–53). The experience of disability identification presents generative opportunities for reforming identification in rhetorical studies and disability studies. Specifically, identification via sensation and touch possesses the potential to reform and reshape the process of identification. For rhetorical studies, the possibilities of touch represent a movement beyond binaries and metaphoric and figurative representations of identification. For disability studies, rhetorical touch offers additional ways of understanding concepts such as community, connection, and interdependence as anchored in lived material realities.

Contemporary disability theorists, attempting to articulate a model of disability that translates across wide audiences of differently disabled people, call for renewed attention to the body in ways that do not create a binary between material and discursive experiences or a dichotomy between individual and social perspectives.[5] This effort often involves a sense of struggle, frustration, and urgency, as theorists attempt to marshal bodily rhetorics to induce potential for identification. As Linton explains, "One research domain that is yet to be fully explored from the perspective of disabled people is the kinesthetic, proprioceptive, sensory, and cognitive experiences of people with an array of disabilities" (*Claiming Disability* 140). Describing her use of a wheelchair in different ways in relation to her body and the world, she explains that "each of these experiences has an impact on my sense of my body in space and affects the information I am exposed to and the way I process sensory information" (140). Linton continues, struggling to articulate the problem in relation to people with disabilities different from hers, such as deafness and cognitive difference, noting how "we are missing the constructs and the theoretical material needed to articulate how impairment shapes disabled people's version of the world. Even as I write this, I am struggling to find the words to describe these phenomena adequately" (140). Linton gropes for words to describe her experience, seeking a model of language *and* of the body that supports her own individual experience of embodiment and that engages with other bodies and disabilities.

Like Linton, Cheryl Marie Wade recognizes the need for an alternative approach to disability discourse and uses blunt language to convey her urgency. She writes, "To put it bluntly—because this need is as blunt as it gets—we must have our asses cleaned after we shit and pee. Or we have others' fingers inserted into our rectums to assist shitting. Or we have tubes of plastic inserted inside us to assist peeing or we have re-routed anuses and pissers so we do it all into bags attached to our bodies. These blunt, crude realities. . . . We rarely talk about these things, and when we do the realities are usually disguised in generic language or gimp humor. . . . If we are ever to be really at home in the world and in our selves, then we must be able to say these things out loud. And we must say them with real language" ("It Ain't Exactly Sexy" 92–93). As Shapiro reports, Wade's plea to talk about these materialities is necessary and urgent. During 504 demonstrations, some protesters "were literally putting their lives on the line, since they risked their health to be without catheters, back-up ventilators, and the attendants who would move them every few hours to prevent bedsores, or who, with their hands, would cleanse impacted bowels every few days" (66–67).

Wade pleads for a language that describes the contradictions between an assumption of personal agency as an autonomous, singular, and independent body and the realities of how some people with disabilities live their lives. This language, in turn, can be marshaled to forge connections and identifications among different people with disabilities and even the larger nondisabled world. She continues, "We who are on the outside, living independently, using attendants for intimate care, owe it to those of our brothers and sisters still dependent on family care or institutions to tell the truth about the pain and struggle of this life as well as the joy and freedom" ("It Ain't Exactly Sexy" 93). She positions the purpose of this communication as a message of informed choice: "So that person in the institution knows that, even with this private need, you have a right to and can set boundaries on who touches you and how you are touched" (93). Wade argues that people with disabilities need to talk about their interdependencies to be at home in the world; conversely, nondisabled people and temporarily able-bodied people need to understand what this world might be like. Beyond a theoretical construct, interdependence is also a "blunt, crude" reality based on the blurring boundaries of bodies. Interdependence necessarily includes touch—the boundaries of how and why people come together. As Wade suggests, there are real material differences that divide people living independently from others in institutions, and yet these differences only make identification more crucial.

Disability occasions a different approach to identification. The metaphoric rhetorics of identification such as those of common ground, interlocking circles, and other mental images do not accommodate many of the disability rights protesters, such as those involved in the demonstrations for 504, who seek to make connections with nondisabled audiences. Metaphorical rhetorics of identification or identification based on binaries of modernism and postmodernism do not accommodate the new models of disability discourse and experience that Linton, Wade, and others imagine. If identification is, as Butler posits, "an assumption of place" (99), then the places where bodies come together in physical contact are crucial. How might rhetorical attention to the sense of touch in these locations foster identification beyond individual disability and among people with varying disabilities, the nondisabled, and the temporarily able-bodied?

Burkean Identification and 504

Touch powerfully catalyzes the values of the disability community, initiating identification and interdependence on material, embodied levels and in symbolic, discursive registers. Attention to touch can also disrupt what Ratcliffe calls the "*false* binary" between modern and postmodern theories of identification (53). Although Burke's concept of identification has often been mobilized to support the ideal of the singular human agent controlling language for his own ends, in opposition to postmodern and discursive processes of identification, Burke's listing of "common sensations" involved in the consubstantiality of identification and his development of the term "substance" propose alternative alignments. Touch, operating as a common sensation in identification, can interrupt the false binary between discourse-based and reality-based determinism and the false opposition between individual agency and discursive production.

Burke's theories of identification elucidate the reasons behind both the successes and the challenges of identification in the 504 demonstrations and can provide opportunities for increasing identification among people with disabilities and their diverse audiences in the aftermath of 504. Burke defines identification capaciously, writing, "You persuade a man only insofar as you can talk his language by speech, gesture, tonality, order, image, attitude, idea, *identifying* your ways with his" (*A Rhetoric of Motives* 55). This process of identifying is intimately connected with the "doctrine of *consubstantiality*," by which "a way of life is an *acting-together;* and in acting together, men have common sensations, concepts, images, ideas, attitudes that make them *consubstantial*" (21). Talking someone else's language, in this sense, stretches broadly to encompass a range of "ways" of life, from speech, tone, and gesture to "common sensations" and other concepts and images that make people connect with each other. In this sense, language—discursive and material, sensory and symbolic—possesses a powerful potential to transform individual and social aspects of experience into material for the process of identification at the collective level.

For people with disabilities participating in 504, identification provided a means by which to unite across individual identities. Assuming that "identity is not merely that which is given to an individual or group, but is also a way of inhabiting, interpreting, and working through, both collectively and individually, an objective social location and group history" (Alcoff 42), rhetorical identification propels a "working through" of identity on both the individual and collective levels. Protesters for 504, for example, possessing a wide range of physical, mental, and cognitive disabilities, did not take it as a given that they did not identify with each other, even as they comprised a "very disparate group, a very wild and divergent community," including, as activist Corbett O'Toole recalls, protesters ranging from "the Mill Valley Moms with a disabled child" to "street junkies" (quoted in Schweik "Lomax's Matrix"). Protesters were able to work with each other, inhabiting a social location based not only through the sit-in of the HEW building but also in the form of a developing group identity. They spoke the same language in a Burkean sense, using common sensations and symbolic identification to resist limitations of individual identity and form a collective.

Rhetorical identification catalyzes the category of identity, enabling connection across identity in symbolic, sensory, and discursive registers. Identity, Siebers explains, "represents the means by which the person, qua individual, comes to join a particular social body" and includes the "capacity to belong to a collective on the basis of not merely biological tendencies but [also] symbolic ones" (*Disability Theory* 15). In other words, identity involves both biological and symbolic elements. In this sense, rhetorical identification via common sensations or touch can transform individual identity for a collective. Eschewing a limited approach to identity in which identity is only "written on the body," demonstrators for 504 enacted a Burkean sense of identification based on a merger of symbolic and biological relationships in both discursive and nondiscursive registers.

Touch activates identification, providing a location for individual and social elements of identity to be realized in sensory and symbolic registers. Protesters for 504 identified with each other through sensory and symbolic ways, establishing a foundation on which potentially to identify with others as well. As the protest and sit-in for 504 demonstrates, disability is an experience that more often than not involves bodies working together, often in partnerships of both disabled and nondisabled people. Protesters for 504 shared "common sensations" in a Burkean sense, initiating identification by acting together and sharing a way of life that conveyed a powerful message about equal rights. Protesters lived together, ate together, moved together, slept together. As Susan Schweik describes in "Lomax's Matrix," demonstrators relied on the Black Panthers and the Butterfly Brigade for assistance in the form of hot meals and walkie-talkies, while making crucial cross-cultural connections. They took care of each other's bodies physically, emotionally, and substantially, living in relation to each other. Using their bodies, they joined their identities together to form identifications with each other and to demonstrate this identification to a largely nondisabled audience.

The Substance of Identification

The common substances and processes of consubstantiality involved in the "common sensations" of rhetorical identification elucidate a means by which touch becomes rhetorical. Touch is a powerful "common sensation" at work in identification, functioning through consubstantiality. In Burke's rhetorical theory, consubstantiality is the merging of "substances" through which one identifies with others (*A Rhetoric of Motives* 21). Burke explains, "A is not identical with his colleague, B. But insofar as their interests are joined, A is *identified* with B. Or he may *identify himself* even when their interests are not joined, if he assumes that they are, or is persuaded to believe so" (20). This process of identification leads to merging of substances: "In being identified with B, A is 'substantially one' with a person other than himself. Yet at the same time he remains unique, an individual locus of motives. Thus he is both joined and separate, at once a distinct substance and consubstantial with another" (21). This locus of substances, at once separate and joined, individual and connective, can be both material and tactile. Substance possesses a materiality comprised of touch that can result in tangible actions: "A doctrine of *consubstantiality,* either explicit or implicit, may be necessary to any way of life. For substance . . . [is] an *act;* and a way of life is an *acting-together;* and in acting together, men have common sensations, concepts, images, ideas, attitudes that make them *consubstantial*" (21). Producing "common sensations," touch, a space both joined and separate, can function in Burkean ways as an "acting together" that leads to identification and consubstantiality. Like Colleagues A and B, demonstrators for 504, many with different disabilities, made themselves consubstantial during the demonstration, and they acted together to reach a larger audience. This acting together included the experience of bodies in contact and relation, as protesters' sensations, concepts, and ideas intermixed in a general way of life over their weeks of living together. When 504 protesters, including their nondisabled allies, engaged in ways of acting together and provided assistance to each other with the necessities of daily life, they initiated processes of consubstantiality.

Extended analysis of the term "substance," applied to the context of the demonstrations for 504, offers possibilities for understanding how demonstrators acted together individually and socially and formed discursive arguments based in material experiences, avoiding the binary of discourse-based and reality-based determinism. In his etymological exploration of substance, Burke draws out its tactile properties, noting that there "is a set of words comprising what we might call the Stance family, for they all derive from a concept of place, or placement" (*A Grammar of Motives* 21). In the Latin and Greek roots of the term "substance," Burke notes a "pun" in that substance often "designate[s] what some thing or agent intrinsically *is*" or "the characteristic and essential components of anything," but writes that it is also a "scenic word" in that "literally, a person's or a thing's sub-stance would be something that stands beneath or supports the person or the thing" (21–22). Burke considers both the literal and metaphorical connotations of "substance" in its etymological roots: "The same structure is present in the corresponding Greek word, *hypostasis,* literally, a standing under: hence anything set under, such as a stand, base, bottom, prop, support, stay; hence metaphorically, that which lies at the bottom of a thing, as the groundwork, subject-matter, argument of a narrative, speech, poem; a starting point, a beginning" (23). Both the materiality and discursiveness of substance, then, contain properties that are connective, even tactile. Inside and outside meet in substance. Substance, "though used to designate something *within* the thing, *intrinsic* to it," also "etymologically refers to something *outside* the thing, *extrinsic* to it" (23). Burke recognizes the tendency to literalize substance but also points toward ways in which its material and symbolic meanings are processed. Intrinsic materiality, for example, can be externalized through discourse.

Relating the importance of this process, Burke writes that "only by systematically dwelling upon the paradoxes of substance could we possibly equip ourselves to guard against the concealment of 'substantialist' thought in schemes overtly designed to avoid it" (*A Grammar of Motives* 57). In other words, thinking through substance is crucial to understanding it and to avoid essentializing it.[6] Furthermore, in regard to substance, "only by learning to recognize its nature *from within* could we hope to detect its many disguises from without" (57). Accordingly, thinking through substance means engaging both internal and external registers. Burke suggests one such method for thinking through substance in his delineation of a type of substance—dialectical substance—through dramatistic analysis, and he notes that in doing so, "we most decidedly do not mean that human motives are confined to the realm of verbal action. We mean rather that the dramatistic analysis of motives has its *point of departure* in the subject of verbal action (in thought, speech, and document)" (33). Consequently the study of substance may take the verbal as one point of departure, but it must also include attention to nonverbal properties, such as the physicality of "stances" and the tactility of the inside and the outside. Identification, then, is material as much as it is discursive and individual as much as it is social and connective.

As Burke explains in *The Philosophy of Literary Form,* "By 'identification' I have in mind this sort of thing: one's material and mental ways of placing oneself as a person in the groups and movements; one's way of sharing vicariously in the rôle of leader or spokesman; formation and change of allegiance . . . the part necessarily played by groups in the expectancies of the individual . . . clothes, uniforms, and their psychological equivalents; one's ways of seeing one's reflection in the social mirror" (227). Beyond verbal analysis, Burke's concepts of identification, substance, and consubstantiality support ways of acting together through a multitude of individual, group, and social "reflections." Similar to how clothes such as uniforms rest on bodies, taking on their shape and shaping bodies as individual but connected, touch, as a powerful common sensation, enacts a physical response and a "psychological equivalent." Touch is a particularly powerful reflection and way of acting together that involves the key ingredients of substance, especially internal and external, material and mental, individual and social, and physical and psychological elements.

The substance of touch propelled identification and ways of acting together among 504 demonstrators. As one demonstrator described, "When one of us was weak, we held the other up" (*The Power of 504*). This sense of support aligns with the Burkean sense of "substance," as it is something that "stands beneath or supports the person" and is "that which lies at the bottom of a thing, as the groundwork, subject-matter, argument" (21, 23). Protesters' demonstrations involved their bodies in material ways, such as in their shared substances in physically caring for each other, while also demonstrating the "substance" or groundwork of the heart of their arguments. As one protester put it, "We're speaking of something beyond our bodies" ("504 Demonstrations"). Another 504 organizer, Kitty Cone, realized that forwarding this argument meant garnering support from organizations "that hadn't thought about disability before," building beyond "individual connections" and thinking of the network of support "as a matrix" (Breslin Oral History 135). Protesters' ways of living together became a model for their larger argument to their largely nondisabled audience. As the activist Mary Lou Breslin described, "The whole thing was like a living role model," explaining that "living out the purpose that you're trying to embody in those regulations—that purpose was being experienced and exercised in the building itself" (Breslin Oral History 123). Protesters supported each other physically and rhetorically, "standing" their ground together during the sit-in and standing their ground in their arguments to a larger audience. Forming a living role model, they laid the groundwork for an argument and for a way of living, demonstrating how identification can be a way of acting together to achieve results. Their individual identities became substantially connected to and supported by each other in material and discursive ways.

Broadcast journalists and protesters frequently used tactile vocabulary, including words employing symbolic and substantial meanings from the Stance family, to describe the protest for 504 and its significance. Heumann, in the early days of the protest, said, "Quite frankly, I think it's going to be very difficult for them to put a lot of pressure on us. When we asked them questions yesterday and we asked them, 'Have you ever read 504?' every one of the people we had in that office said, 'No'" (*The Power of 504*). Heumann described this situation as an opportunity for people in the HEW offices to read 504 and "get educated about law that they're supposed to be enforcing" (*The Power of 504*). Putting the situation in terms of relative pressure and counterpressure, Heumann used the office's lack of knowledge as a way to exert pressure on them. Soon after this, journalists described the HEW officials as "stiffening their easy attitude about the demonstrators" after the second night of the occupation (*The Power of 504*). Two weeks into the occupation, a newscaster described a rally outside the HEW building as offering "the mental and vocal backing that those inside really needed" (*The Power of 504*). In Burkean terms, this "backing" worked as a "stand," "support," or "base" substantially and discursively for the protesters engaged in the sit-in inside and the outside rally, exerting pressure both intrinsically and extrinsically on the HEW to listen.

The paradoxes and ambiguities of substance and the words of the Stance family are particularly abundant in the notion of 504 protesters standing their ground and standing by their arguments in their occupation of the HEW building. Some protesters could not physically stand on their own; some protesters with cognitive disabilities were not able to make traditionally persuasive verbal or written arguments. But protesters joined their bodies, minds, and all their ways of being to act together during the sit-in. As Wayne Booth describes in "The Rhetorical Stance," the stances that do not fit the norm of communication models are often the ones that are most describable. Exploring various "perversions" of the rhetorical stance and in search of a stance of "balance," he notes that "balance itself is always harder to describe than the clumsy poses that result when it is destroyed. But we all experience balance whenever we find an author who succeeds in changing our minds. He can do so only if he knows more about the subject than we do, and if he then engages us in the process of thinking—and feeling—it through" (144–45). Protesters for 504 were particularly successful in their rhetorical stance, achieving success with a demonstration of traditional balance by changing minds and teaching audiences but also by using nonnormative poses and differently balanced arguments based on rhetorical touch, which encouraged a "feeling through" of their position. By supporting each other, they made use of the complexities of substance and stance, holding each other and themselves up literally, discursively, individually, and socially.

Delimiting and Intertwining Touch

In spite of false binaries that would pit modern and postmodern forms of identification against each other, Burke's tactile notion of substance and stance supports an understanding of identification suffused with both modern and postmodern elements. Similarly, postmodern theories of identification, especially ones that involve touch, are saturated with modern elements of identification. Tracking elements of Burkean identification through phenomenological, deconstructive, and postmodern theories of touch reveals a theory of identification via sensation that eschews easy binaries. The philosophers Nancy and Merleau-Ponty extend Burke's concept of identification, offering models of embodiment that position the body in constant relation to other bodies and suggest alternatives for identification across the individual/social and material/discursive binary, particularly in contexts of disability. Both philosophers turn to the sense of touch to pose and respond to questions regarding the body's limits when they experience disability or bodily vulnerability. These questions and contingent answers that touch provokes constitute additional ways of rereading protests such as 504 into contemporary remembrances and for potential future action.

Touch characterizes writing in general for Nancy, providing a point of entry for understanding the individual body in relation to the social body, in material experience and in discourse. For Nancy, "Writing touches upon bodies *along the absolute limit* separating the sense of the one from the skin and nerves of the other" (11). He claims to "*know of no writing that doesn't touch.* Because then it wouldn't be writing, just reporting or summarizing. Writing in its essences touches upon the body" (11). Like Burke's invocation of the literal and symbolic meanings of "substance," this touch yokes the literal, material, and sensory aspects of the body with other bodies in connection with the figural, discursive, and rhetorical aspects of writing. The limit denotes for Nancy the location where these registers meet: "*Bodies don't take place in discourse or in matter.* They don't inhabit 'mind' or 'body.' They take place at the limit, *qua limit:* limit—external border, the fracture and intersection of anything foreign in a continuum of sense, a continuum of matter" (17). This limit, material and metaphorical, is where discourse and matter meet, as a writer "touches by way of addressing himself, sending himself *to* the touch of something outside, hidden, displaced, spaced . . . remaining a stranger *to* contact *in* contact: that's the whole point about touching, the touch of bodies . . . the writing 'I' is being sent from bodies to bodies" (17, 19).

Touch, especially in its function as an address to both oneself and one's audience, is rhetorical for Nancy in multiple ways. Bodies "take place" materially and discursively at the site of touch; all writing is tactile to Nancy. The "touch of bodies" is equivalent to the "writing 'I'" sent among bodies in contact, initiating identification materially and rhetorically beyond the individual. Touch delimits bodies and invites identification within self and audience, blurring the boundaries of the traditional rhetorical situation that may separate the bodies of writer from reader, addresser from addressee, or rhetor from audience.

For the phenomenologist Merleau-Ponty, touch also forms a site for connection, both material and discursive, between bodies. Merleau-Ponty turns to the experience of a "reversible handshake" to explore this relationship: "If my left hand is touching my right hand, and I should suddenly wish to apprehend with my right hand the work of my left hand as it touches, this reflection of the body upon itself always miscarries at the last moment: the moment I feel my left hand with my right hand, I correspondingly cease touching my right hand with my left hand" (9). In this paradox of sensation, Merleau-Ponty cannot ascertain which hand is doing the touching and which hand is feeling the touch. Both the inside and outside of the body produce sensation in a reversible handshake and invite reconsiderations not only of the boundaries of his own body and sense of self, but also of the boundaries between himself and others. Whereas Nancy uses the concept of the limit, Merleau-Ponty uses the rhetorical figure of the chiasm, or *chiasmus,* from the Greek meaning "to shape like the letter X," to describe the reversibility of his handshake. This figure often refers to two clauses that reverse in structure, but more generally it means any intertwining of ideas in a crisscross configuration. Merleau-Ponty's reversible handshake, with hand touching hand, initiates this rhetorical device and invites the possibility of identification across bodies beyond his own.

As with Nancy's probing of the limit, Merleau-Ponty's experience of the indeterminacy of the touching and touched hands reflects a larger effort to think outside of the self. Merleau-Ponty asks, "What if I took not only my own views of myself into account but also the other's views of himself and of me?" (8). He realizes that the paradox of sensation he feels in his reversible handshake, which straddles both internal and external senses of his body, is applicable to other bodies as well: "Why would this generality, which constitutes the unity of my body, not open it to other bodies?" (142). The double sensation that Merleau-Ponty experiences within himself in his reversible handshake connects him to other bodies, moving him to an understanding of his body as primarily in relation to other bodies. Sensation is already doubled in himself and becomes increasingly replicated among other bodies as he realizes the inherent connections between individual and other bodies via touch.

Touch—conceptualized as the limit for Nancy and the *chiasmus* for Merleau-Ponty—embodies Burkean notions of the biological and symbolic processes of identification, including attention to both intrinsic and extrinsic experience in material and discursive registers. Touch for both Nancy and Merleau-Ponty is a common sensation that works rhetorically to invite identification. Touch, in Burkean terms of identification, exemplifies the feelings of being "both joined and separate, at once a distinct substance and consubstantial with another" (*Rhetoric* 21). Touch also creates a "substance" and forms a "stance" for each philosopher, exhibiting Burkean material-discursive properties. The descriptions of the limit for Nancy and the *chiasmus* for Merleau-Ponty operate as the metaphoric "groundwork" or base of their arguments but also indicate the physical substance of discourse. For Merleau-Ponty, especially, the reversible handshake, with hands supporting and "standing under" one another, operates as a model for thinking how all bodies support each other. In Burkean terms, the hands are literally "something that stands beneath or supports" and operate metaphorically, in Merleau-Ponty's description, as the "groundwork" or base of an argument (Burke *A Grammar of Motives* 22–23). For languages that take place primarily via the hands, such as American Sign Language, or, to a certain extent, facilitated communication, communication operates in this material-discursive register in literal and symbolic ways.

For both Nancy and Merleau-Ponty, the experience of bodily vulnerability and disability further positions the sense of touch as a way of blurring boundaries between oneself and others, in writing and in experience. When Nancy experiences a particular bodily vulnerability—a heart transplant—his conceptualization of touch as a material-discursive limit is animated by real life experience. Nancy struggles at first to define the difference he feels as a result of his failing heart, wondering whether to identify himself as "disabled," "ill," or simply aging (163). He settles on "ill" but admits, "Ill is not exactly the term: not infected, just rusty, tight, blocked" (163). Contemplating his heart, Nancy remarks that "up until this point, it was strange by virtue of not being even perceptible," and yet because he is ill, his heart "has to be extruded" and becomes strange again—"intrusive"—when he receives the new donor heart (163). He muses, exploring the limits of his idea of identity: "I (who, 'I'? this is precisely the question, the old question: who is the subject of this utterance, ever alien to the subject of its statement, whose intruder it certainly is, though certainly also its motor, its clutch, or its heart)—I, then, received someone else's heart, about ten years ago. It was grafted into me. My own heart (you will have understood that this is the whole question of the 'proper' . . .)—my own heart, then, was useless for reasons never explained. In order, therefore, to live, I had to receive the heart of another person" (162). Nancy's identity is indeterminate after the transplant, an indeterminacy that he realizes has always been there.[7] He describes the strangeness of another's heart grafted into him as a "strangeness [that] binds me to myself" (163). The transplant, therefore, at once creates distance and connection in Nancy's self-identification. He wrestles with a failing heart that he can no longer call his own while he also struggles to call the donor's transplanted heart his own. This strangeness begins with the sense of touch—what he calls the "physical sensation of a void" in his chest, which subsequently includes many additional touches and sites of material contact—the incision, the "opening up the entire thorax, taking care of the graft-organ, circulating the blood outside the body, [and] suturing the vessels" (162, 165). For Nancy, these sites of touch embody the paradoxes of substance, at once dissociating him from himself and, eventually, opening him up to others.

Like Merleau-Ponty contemplating his reversible handshake, Nancy identifies himself as "doubled" after his transplant. Nancy also identifies the people to whom he is connected via the transplant as doubled and even multiplied. Nancy's questioning of his use of "I" leads him to realize that the transplant—the site of so many intrusive touches—also connects him to "several others," including, predictably, those who are close to him but also the doctors and other medical, legal, and social personnel, who, like himself, he realizes are also "now doubled or multiplied more than ever before" (163). Picturing the complexity of "this strange group," he wonders about questions of exactness and justice, including whose heart he possesses and how he was chosen to receive it (163). He defers these answers at first and then settles on an assessment that bridges internal and external registers: "This whole thing will reach me from somewhere else and from outside—just as my heart, my body, are reaching me from somewhere else, are a somewhere else 'within' me" (164–65).

Beyond the experience of the initial intrusive touch—the transplant procedure—Nancy involves another set of contacts and connections. In particular he notes the "great emphasis" placed on "the solidarity, and even the fraternity, of 'donors' and recipients" in the transplant process (166). In fact his meditation on touch, especially its material and discursive properties, leads him to remark on the delimiting potential of organ donation, as he states that "freed from any limits other than blood-group incompatibility (and freed especially from any ethnic or sexual limits: my heart can be a black woman's heart)—that this gift institutes the possibility of a network where life/death is shared by everyone, where life is connected with death, where the incommunicable is in communication" (166). Connecting his individual experience to a network of others, many of whom are doubled and multiplied like him, enables Nancy finally to accept the strangeness of the new sensations that accompany the heart transplant. He writes, "Thus, then, in all these accumulated and opposing ways, my self becomes my intruder" (168). He continues, "I certainly feel it, and it's much stronger than a sensation: never has the strangeness of my own identity, which for me has always been nonetheless so vivid, touched me with such acuity. 'I' clearly became the formal index of an unverifiable and impalpable change. Between me and me, there had always been some space-time: but now there's an incision's opening, and the irreconcilability of a compromised immune system" (168).

Nancy's understanding of touch becomes more far-reaching after this realization. The sense of touch opens up a range of meanings that double and multiply, connecting Nancy's singular experience of an intrusive touch—the incision—to a wider network of figurative and literal connections through which he is in touch with and touched by others. Nancy reports a "general sense of being no longer dissociable from a network of measures and observations—of chemical, institutional, and symbolic connections that do not allow themselves to be ignored" (169). He relates, "The empty identity of the 'I' can no longer rely on its simple adequation (in its 'I=I') as enunciated: 'I suffer' implicates two I's, strangers to one another (but touching each other)" (169). Nancy feels the weight of these multiple I's with tactile acuity, in writing and in his daily life, a feeling that is more powerful than sensation because it initiates an identity intimately connected with other audiences and networks. His increased perceptions of his heart lead directly to his positioning of himself in these groups, a position in which Nancy uses literal and discursive touch to mediate between social and individual identities. Both chemically and symbolically, he moves from the singular "I" of identity to a multiple "I" of identification with others in a living network. In Burkean terms, Nancy's experience of common sensations leads to a merging of identity with identification.

A form of bodily vulnerability also causes Merleau-Ponty to prod the limits of identity and to move toward identification. Considering the sense of touch in relation to vulnerability, Merleau-Ponty proposes the question "Yes or no: do we have a body—that is, not a permanent object of thought, but a flesh that suffers when it is wounded, hands that touch?" (137). He answers this question by noting that any response that ignores the literal properties of the body—its aptitude for touch and its vulnerability—would lead "to the bifurcation of subject and object" (137). Merleau-Ponty draws again on the figure of the *chiasmus* to conceptualize how touch unites bodies in language and the material world. This rhetorical figure of intertwining, a literal crisscrossing of symbols and ideas, means that language literally turns back on itself, producing reflection and reordering, blurring subjects and objects. Returning to the sensations produced by his hands touching each other in the *chiasmus* of his reversible handshake, Merleau-Ponty considers this experience as the "initiation to and the opening upon a tactile world" (133). Language and communication are key elements of this tactile world. At once material and discursive, in body and in language, this opening up of a tactile world is based on social and individual connection.

Recalling his reversible handshake, Merleau-Ponty explains, "There is a reversibility of the seeing and the visible, and as at the point where the two metamorphoses cross what we call perception is born, so also there is a reversibility of the speech and what it signifies; the signification is what comes to seal, to close, to gather up the multiplicity of the physical, physiological, linguistic means of elocution, to contract them into one sole act" (154). This one act contains multiplicities, suggesting possibilities beyond the discrete elements usually understood in speech and signification—the rhetor, the audience, and the message. Although Merleau-Ponty is concerned with vision as well as touch, it is the reversibility in each of these senses that binds one to another and that provides a model for understanding language and its elements as flexible media. The *chiasmus* is more than simply a figure of speech; instead it is an embodiment of the relationship between language and bodies, not necessarily a synthesis but a system of connection and reversibility in which bodies can remain both connected and separate. Discourse and materiality, individuals and social bodies are connected in the *chiasmus.*

Nancy's and Merleau-Ponty's philosophies of touch offer generative models for the kind of identification across diverse bodies that people with disabilities such as Heumann, Linton, Wade, and others seek.[8] The solidarity of donors and recipients involved in Nancy's transplant is a living network not unlike the living matrix of the 504 demonstration. The *chiasmus,* especially as it functions in Merleau-Ponty's physical experience of the reversible handshake and his discursive experience as a model for using language to connect oneself to others, provides opportunities for identification, reflecting the experiences enacted by the 504 demonstrators. As he identifies himself with a tactile world, Merleau-Ponty connects his experience, in body and in language, to others. Likewise, Nancy's concept of the limit as a space for touch in writing and in bodily sensation opens up to him possibilities for re-identifying with himself and with the various other people and social networks involved in his transplant. The *chiasmus* and the limit provide models for understanding touch as a way of moving from an understanding of embodiment as based in the singular body to seeing embodiment as based among interdependent bodies in contact. This more flexible approach to embodiment, especially the reversibility of the *chiasmus* and the connection of the limit, offers a model for embodiment to take place in a space that resides neither in self nor other, inside nor outside, body nor discourse only. This intertwining and delimiting of touch among bodies ensures connection, permeability, and even separation among bodies in identification.

Common sensations occasioned by disability or bodily vulnerability also initiate identification and blur the separations between bodies often assumed by traditional approaches to rhetorical situations. Merleau-Ponty and Nancy model an identification that takes place in a third space, beyond the binaries of the body as social/individual and material/discursive. They enact a Burkean kind of rhetorical identification through touch in which inside and outside, symbolic and biological, and mental and material elements intertwine. Each philosopher relies on his sense of individual bodily experience as a force in shaping his approach to language, but each also relies on a sense of social and discursive connection through touch to bridge differences and to create common ground among different audiences, some with different interests. Neither positions his experience as simply a product of various forces of discourse; nor is either completely in control of his discourse. Nancy and Merleau-Ponty conceive of their experiences not as discourse-determined or reality-determined but instead as third spaces where these identifications play out across bodies and rhetorics. This self, redefined through the sense of touch, is primed for identification. Touch disrupts a sense of control over the singular body or language and instead replaces it with a nexus of new possibilities for identification.

Both Merleau-Ponty and Nancy illustrate, in Burkean terms, ways of life that constitute "acting together," in which the substances and sensations of touch intermingle bodies with discourse and push the limits of identification and identity. Nancy, in particular, becomes consubstantial with not only the donor of his heart but also the wider network of people and institutions involved in his transplant. He continues to work together with these people and networks in his own writing, memories, and descriptions of the transplant event. Merleau-Ponty uses the prospect of bodily vulnerability to imagine connections among people in action and identification. Both philosophers set up their experiences with disability as models for understanding how all bodies act together to forge new areas of identification.

Folding Identifications

The intensity of contemporary visual imagery, Constance Classen notes, might make touch "the hungriest sense of postmodernity," but touch has been taken up "at least edgewise" by a number of philosophers (*The Book of Touch* 2). French philosophers of the late twentieth century, for example, such as Jacques Derrida, Gilles Deleuze, Felix Guattari, Luce Irigaray, and Hélène Cixous, explore touch in relation to self, society, gender, and language. As Merleau-Ponty does with *chiasmus* and Nancy with limit, Deleuze uses the sense of touch through the concept of the fold to disturb traditional oppositions and to locate new spaces for identification. The fold, when understood in relation to spaces of struggle, offers new ways of acting together that facilitate identification.

Deleuze uses the fold to understand and organize baroque philosophy, finding it present in everything from theories of matter and bodies in the emerging sciences to styles of dress and architecture. In relation to the body, the fold invites flexibility. Like the limit and the *chiasmus,* the fold is a way of embodying both sameness and difference and self and other in locations that invite alternative identifications. Tracing the concept of the fold, Deleuze relates, "The fold can be recognized first of all in the textile model of the kind implied by garments: fabric or clothing has to free its own folds from its usual subordination to the finite body it covers" (*The Fold* 121). This freedom results in "the fold that goes out to infinity" in which the "thousand folds of garments . . . tend to become one with their respective wearers, to exceed their attitudes, to overcome their bodily contradictions" (121–22). The fold overcomes bodily contradictions while simultaneously producing a sense of cohesion. As Deleuze, illustrating the concept with the work of philosopher Gottfried Leibniz, describes, "A flexible or an elastic body still has coherent parts that form a fold, such that they are not separated into parts of parts but are rather divided to infinity in smaller and smaller folds that always retain a certain cohesion" (*The Fold* 6). In other words, bodies are both continuous and differentiated; cohesion may bind a body, but within that cohesion resides infinite difference.[9] In Leibniz's example, "the division of the continuous" is similar to the effect of "that of a sheet of paper or of a tunic in folds, in such a way that an infinite number of folds can be produced, some smaller than others, but without the body ever dissolving into points or minima" (6).[10] Deleuze uses the example of the folded paper as a way to think visually about the fold, particularly about how it changes the surface, shape, and direction of the paper.

The fold, as a tactile concept, can also provide a model of "acting together" with other bodies, through folds in space, time, and thought. The "fluid" properties of the fold, for example, are also present in Deleuze and Guattari's notion of smooth space as opposed to striated space. The smooth space is a liquid, haptic space, which is aligned with an intense, close vision that enables difference and variation. They relate, "The first aspect of the haptic, smooth space of close vision is that its orientations, landmarks, and linkages are in continuous variation" (*A Thousand Plateaus* 493). In Deleuze and Guattari's aesthetic model of art, this variation, a result of intense relation between touch, vision, and writing, forms a locus of possibility. Similar to Merleau-Ponty's *chiasmus* and Nancy's limit, the fold is a tactile model for uniting material and discursive and individual and social approaches to identification. The haptics of smooth spaces are places in which "the struggle is changed . . . life reconstitutes its stakes, confronts new obstacles, invents new paces, switches adversaries" (500). Although Deleuze and Guattari note that these spaces are not necessarily liberatory, they offer locations in which change is possible.

The fold, as it operates as an interruption in assumptions regarding the "normal" sense of space, when applied to the issue of disability, offers the possibility of reconfiguring traditional notions of ability and disability. In Burkean terms, the fold can be understood to support a way of acting together that includes identification as well as division. The fold, enveloping both continuity and differentiation in bodies, represents the myriad ways in which ability and disability are parts of the same overarching system. One potential liberatory space is the memorial of the 504 demonstration, particularly how people with disabilities continue to fold the demonstration for 504 into their lives, rewriting and reshaping this event into the present and future, increasing opportunities for identification among people with and without disabilities.

Memorials of the 504 sit-in are a telling demonstration of the rhetorical power of the sense of touch and its potential for new spaces of identification. In her reflection on the protest, Joan Tollifson, a 504 protester, compares her life, including her approach to disability, before and after the demonstration. She describes herself before the demonstration as angry and violent, drinking excessively, consumed by drugs, and a frequent visitor of jails and hospitals. She purposely avoids identification with disability, writing, "Growing up, I wanted to dis-identify myself with the image and label of being a cripple. I wanted to be normal . . . I avoided other disabled people. I refused to see myself as part of that group" (106). After a particularly low point in her life, she found a therapist who recommended joining a group of disabled women. Tollifson says that she resisted the idea at first but came to see this experience as a key element in turning her life around. She writes of the group, "They shared so many of what I had always thought were my own isolated, personal experiences that I began to realize that my supposedly private hell was a social phenomenon" (106–7). She details this realization: "We had eye-opening, healing conversations. We discovered, for example, that we had all had the experience of being patronized and treated like children even though we were adults. It wasn't simply some horrible flaw in my own character that had provoked such reactions, as I had always believed, but rather, this was part of a collective pattern that was much larger than any one of us. It was a stereotype that existed in the culture at large. Suddenly disability became not just my personal problem, but a social and political issue, as well" (107).

Tollifson turned her experience of disability inside out, folding it in with the experiences of other women with disabilities individually and socially and materially and rhetorically. Moving from her own, isolated personal experience to the experiences with the group, she began to identify herself with them and found common ground and identification in the discrimination she had experienced. With these women, she took part in the month-long occupation of the San Francisco federal building for 504. She describes activities ranging from wheelchair races to strategy meetings, as people "laughed, argued, [and] shared their lives" together and found herself feeling, for the first time, "like a real adult member of the human community" (107). This experience united her individual and social identity and delimited potential spaces for identification: "This revealed exactly how disability colored my life (in the way the world saw me, and inside my own head as well)" (107). In her writing and in her world, Tollifson began to identify with other people with disabilities. Rhetorically, in Burkean terms, the 504 demonstration became for Tollifson a space in which substances such as the common sensations, attitudes, and ideas that protesters shared were acts and a "way of life" that functioned as "an *acting-together*" (21). Within the space and activities involved in the 504 demonstration, Tollifson began to experience consubstantiality and then identification.

She enacted Nancy's concept of the limit and Merleau-Ponty's *chiasmus* in her life after the demonstration, using touch to fold new identifications into her interactions with a wide range of people, both disabled and not.

Tollifson's new approach to the sense of touch after the 504 demonstration clearly shows how she turned her individual and social identity inside out, crisscrossing and exploring the limits of each. After the sit-in she began seeking out new experiences. She writes, "In the years after the 504 sit-in, I took up karate and broke boards with my arm. I studied massage and began doing it for a living, breaking another taboo. You aren't supposed to touch people with a 'deformed' body part; it may be disgusting to them. That deep inner feeling of being disgusting still resurfaces every time I touch someone for the first time with my arm. The healing occurs slowly, over a lifetime" (107). Touch became for Tollifson a way of interacting with the world. She initiated consubstantiality with people through the tactile properties and sensations involved with karate and massage. After taking up karate and massage, both activities that require her to use her arm differently, Tollifson became more creative with touch. When studying meditation, for example, she was at first dismayed that she could not make the traditional Zen mudra that requires that hands are touching, because she was born without an arm. Eventually she realized that her body's unique shape could form an alternative mudra. She embodies the concept of this fold in this exercise, inhabiting a sense of cohesion in difference. Her new shape at once "overcomes bodily contradictions" in the spirit of the concept of the fold and also opens Tollifson's body and mind up to accepting and valuing a wider range of bodily shapes and embodied practices, without the telos of overcoming. Through transformations such as these, she also began to move from thinking of herself as a body and toward reorientating herself among bodies in contact and connection.[11]

New ways of coming in contact with her body and with others also enabled new rhetorical connections for Tollifson. Through connecting with the disabled community and participating in events such as the sit-in and activities such as karate, massage, and meditation, Tollifson learned that she can connect with people in new ways, both literally and rhetorically. In the disabled community she realized that "here was a society where being disabled was no big deal" (107). This is in stark contrast to the encounters in an ableist world that she experienced, in which, as she describes, the "question of 'what happened to your arm?' has followed me through life like some koan-mantra that the universe never stops posing" (105). Tollifson reports receiving a wide range of responses and unsolicited advice regarding her missing arm—people cry or tell her she is an inspiration, children gawk and ask questions, and others pretend desperately not to notice. After experiencing life within the disabled community, Tollifson took a matter-of-fact perspective on her disability, writing, "I'm missing my right hand and half of my right arm. They were amputated in the uterus, before I was born, by a floating fiber" (105). The disability community has shown her that the answer to the question of what happened to her arm is to form new connections with her arm—to break boards with it, to form meditations with it, and to massage. Her experience at the 504 sit-in made her consubstantial with the disability community, and she has carried this consubstantiality over into her relationships with nondisabled people as well.

Tollifson folded her experience of sitting-in for 504 back into her life and in the decades that followed, returning to her experience at the demonstration to establish new contact, materially and rhetorically, with the world that facilitates new opportunities for identifications. Reflecting on rhetorical encounters with other people with different disabilities, she writes, "One can look at a quadriplegic or amputee and feel pity, or in another moment, one can feel attraction and envy, as I had discovered during the 504 sit-in. . . . Anything is possible" (111). She relates, "In a way I've gone back to the innocence of a baby. When little babies encounter my arm—the arm that extends just below the elbow—it is seen as just another interesting shape to explore" (111). As an adult she has found a way of going back to that innocence through meditating, which she calls a "listening stillness" and "open seeing" (111). This sensory experience, however, goes beyond the eye and the ear. Tollifson, in touch with the world through her arm, has established a method for understanding disability from a perspective that is skeptical of binaries and contradictions. Tollifson uses touch and a concept similar to the fold not so much to overcome bodily contradictions as to demonstrate that overcoming is not necessary and that contradictions may not exist, as all bodies are connected through difference and continuity. Through touch she avoids "all the conditioned concepts and judgments that are superimposed by our thinking"; practices "not labeling what appears as either a deficit or an asset, perfect or imperfect, beautiful or ugly"; and finds a space where she can explore "wondering openly, without conclusions, without trying to get somewhere else" (111). Like the space of touch described by Merleau-Ponty and Nancy, a third space has been established by Tollifson via touch, a place in which she can fold and refold her experience of the 504 sit-in into new identifications with both disabled and nondisabled people.

This third space has enabled Tollifson to revise antagonistic assumptions about rhetorical encounters based on opposition between rhetors and audiences or assumptions about the gulf of identification between disabled and nondisabled people. Reevaluating the traditionally agonistic rhetorical encounter, she writes, "We can perhaps begin to see and question the strong tendency to polarize and create enemies to blame and hate, the tendency to defend our own opinions as if our life depended on them. We become less identified with our ideas. We develop a greater ability to listen to differing concerns, to notice how we become attached and upset" (111–12). By seeking ways to become less identified with her ideas, Tollifson has found ways to be more receptive to alternative, even opposing ideas, inviting the possibility to become more identified with different types of disabled and nondisabled people. Touch invites a different approach to rhetorical encounters, enabling open-mindedness and strategies of listening in which distinct boundaries between the entrenched positions of audience and speaker are blurred.[12] Both identification and division function together in this process, as each, in Burkean terms, works toward an invitation to a rhetoric based on openness and difference.

Touch and Online Memorials of 504

Within the disability community and its allies, touch can function as a valuable strategy for ensuring connection and coalition-building in the face of continuing challenges and efforts to isolate and individualize. As Anne Finger writes, the "notion that disabled people should be alone, isolated, not in community with others —whether nondisabled or disabled—is [still] deeply entrenched" in the cultural consciousness, so much so that the "idea of a disabled person in relation with another person is unsettling; we can scarcely bear to think about it" (611). Burkean approaches to substance, in conversation with phenomenological and deconstructive approaches from Merleau-Ponty, Nancy, and Deleuze and Guattari, offer a way to think about connection and relationship in the context of disability. The fold, the *chiasmus,* and the limit offer ways of understanding bodily substances and discourses in relations of interdependence and contingency. Valuing these offerings is crucial for the maintenance of the disability community and its connections to nondisabled allies; as Finger warns, "Physical segregation of disabled people from the broader community may be waning, but psychic segregation seems to be alive and well and living on the fringes of our consciousness" (611). In one form of resistance to this psychic segregation, disabled people and their allies mobilize memories of 504 and its successes.

The remembrance of 504 is a form of activism that a wide range of people fold into their approaches to disability rights, embodiment, and identification, continuing to prod the clear lines between "us" and "them" or "rhetor" and "audience," especially in online registers. The twentieth and thirtieth anniversaries of the 504 demonstration have been commemorated online in spaces that invite an active form of remembering in which communicators join with audiences to fold the demonstration into their present and future goals. One online writer of a nondisability-focused newspaper chose to return to the physical and bodily exertions demanded by the sit-in, urging nondisabled readers to try to feel what the demonstrators must have felt: "Imagine what it would be like if you depended on a wheelchair for mobility and the only bathroom on the floor had doors so heavy that they must be propped open for easy entrance. Imagine, for that matter, what it was like for those who typically depended on personal assistance to dress and bathe to stay in one place. All for a principle" (Kendrick). Like the demonstrators in 1977 who used powerful bodily rhetorics to get their message across, remembrances of 504 emphasize bodily experience to increase identification between writer and reader. Remembrances such as these create a haptic space of liberatory potential through increased identification and blurred boundaries between rhetors and audiences.

Other online memorials, such as those maintained by the Disability Rights Education and Defense Fund, focus on folding into the memorial actual footage of the demonstration and reprinted testimonials with video recollections made by participants twenty years later, thereby making this material more available to online disability writers and bloggers ("504 Sit-in 20th Anniversary"). The online archive of 504, including a video documentary available on YouTube, inspires commentary in the disability blogging world. The blog *The Gimp Parade*, for example, reprinted Ed Roberts's and Judy Heumann's testimonies to the congressional committee, splicing into them links to background information on the rehabilitation act and disability rights history for new readers and people new to the movement. Blogger Kay Olson places her present daily life and recent disability rights history into the events of 504. Olson, who uses her blog to connect to a wide range of other disability bloggers, writes of 504: "The ADA was possible because of this. My education and ability to sit here and type today was [*sic*] profoundly effected [*sic*] by the actions of these disability rights heroes of the past. Just thirty years ago" ("504 Sit-in 20th Anniversary"). A commenter, remarking on Olson's post, writes, "I'm in San Francisco today. . . . While I was wandering around downtown I reminded myself that it was thirty years since that amazing moment in the city's and the whole national movement's history. It's a story that deserves to be more widely remembered—loads of dramatic potential, too—instead of more 'inspirational' movies about pwds [people with disabilities], I'd like to see '504,' coming soon to theatres near me" ("504 Sit-in 20th Anniversary"). Olson and her commenter draw from everyday, embodied experiences—typing on a keyboard, moving through a city—in their reflections and appreciations of the demonstration. Online acts in potentially liberatory spaces such as these fold into the 504 memorials the types of activities and rhetorics that keep the movement going.

The sit-in for 504 and its memorial show how it is possible to reimagine identification in the context of disability. The sense of touch, especially as it provides a third space for resisting the binaries of the body as individual or social, material or discursive, disabled or nondisabled, moves writers such as Tollifson toward a connective approach across wide audiences. In online communities, this tactile approach is folded in through remembrances and testimonies of other people's lives affected and changed by the events of 504. Rhetorically touch provides a space for rhetors to listen and react receptively, making connections at the individual and social levels in embodied and discursive ways. Touch delimits, intertwines, and folds the singular elements of rhetorical situations into opportunities for rhetorical identifications among bodies in contact, blurring boundaries between rhetor and audience. Accordingly the following three chapters take specific elements of traditional rhetoric—*logos, ethos,* and *pathos*—and use the tactile rhetorical practices involved in Empedocles's felt *logos, mētis,* and *kairos* to explore more fully the identifications and possibilities afforded by folding the practices of touch into traditional and contemporary rhetoric.

3

Feeling *Logos*

Empedocles's Repetitive Rhetoric and Psychological Disability

Identification can be a challenge for people with a wide range of disabilities, but people with psychological disabilities may face especially difficult obstacles communicating messages and initiating identification with audiences, especially nondisabled ones. People with psychological disabilities are often prone to the accusation that they lack the appeal of *logos*, a particularly harmful charge because *logos* is often considered the appeal most characteristic of Aristotle's rhetoric and is a key element in rhetorical connection and identification with others. *Logos*, as it is traditionally understood, is the message or argument of a rhetor's speech or writing and is usually associated with reason, logic, and rationality.[1] The communications of people with psychological disabilities are frequently designated as lacking reason, logic, and rationality. The language of people with schizophrenia, for example, is commonly characterized as "word salad" or "verbal diarrhea," and the language of anybody with a psychological disability—from psychosis to depression—is often doubted as having any rational or logical consistency.

To theorize the rhetorics of people with psychological and mental disabilities, rhetoricians such as Catherine Prendergast, Cynthia Lewiecki-Wilson, and Margaret Price turn to alternative models of rhetorical agency. Exploring the "rhetoricability" of people with schizophrenia specifically, Prendergast wonders how the speech and writing of people with mental illness can ever become anything more than a "rhetorical black hole," a designation that avoids the crucial question of "how or maybe when does one honor schizophrenics as rhetorically enabled subjects?" ("On the Rhetorics of Mental Disability" 56, 54). Exploring the "rhetoricity" of people with severe communicative impairments who may lack the ability to speak or write, Lewiecki-Wilson suggests a model of facilitated or mediated rhetoric, via Krista Ratcliffe's model of "rhetorical listening." Rhetorical listening for people with mental or psychological disabilities "could be understood as thoughtful attention to the disabled person's sounds, habits, moods, gestures, likes and dislikes" as well as "the disabled person's embodied, nonverbal performances and preferences for daily living" (Lewiecki-Wilson "Rethinking Rhetoric" 161–62). Similarly, Price seeks ways of "listening to the subject of mental disability" that draw on alternative understandings of "reason" and "criticality" (*Mad at School*

25, 30). The stakes of this effort are extremely high for people with psychological disabilities because existing theories of rhetoric, even those offering revisions to dominant paradigms, are inadequate. As Price states, "Despite many attempts to refigure rhetoric through multiracial, feminist, poststructural, sophistic, and other theories, the underlying presumption of reason-as-normality has remained largely unchanged" (32). The assumption between reason and normality is particularly trenchant in relation to the appeal of *logos* and continues to exclude rhetors with psychological differences.

Attending to the sense of touch is another way of valuing the rhetoricity and rhetoricability of people with psychological disabilities and of facilitating potential rhetorical identification. As is explored in this chapter, people with psychological disabilities, ranging from schizophrenia to depression, use touch to communicate with others and to convey a sense of felt *logos* in their rhetorical productions. These communications operate beyond traditional understandings of logic, reason, or rationality, forming identifications based on mutual ways of feeling and experiencing relationships and the world. Theorizing a felt sense of *logos* among people with psychological disabilities enables the possibility of identification with diverse audiences outside of narrowly rationalist paradigms.

Extending the work of Prendergast, Lewiecki-Wilson, and Price, who seek to find alternative models for valuing the rhetorics of people with psychological disabilities, I offer an alternative definition of *logos* based on the pre-Socratic philosopher Empedocles's teachings and theories, which emphasize the potential tactile and nonrational properties of *logos.* After exploring the challenges that people with psychological disabilities undertake in communicating their messages, I reread Empedocles's theory of *logos* as a performance of *logos* based on proportions of physicality and reciprocity between bodies and minds, arguing that it models a felt sense of *logos* that values diverse psychologies and embodiments that invite identification. I also apply Empedocles's performance of *logos* to the real-life experiences and online writing of people with psychological disabilities who use the sense of touch as a rhetorical strategy to create identifications among themselves and other people. Exploring these tactile rhetorical strategies simultaneously revises rationalist definitions of *logos* for rhetors with psychological disabilities and expands opportunities for identification among wide ranges of people—disabled, nondisabled, or temporarily able-bodied.

Logos and Psychological Disability

In the *New York Times* best seller *The Soloist,* the journalist Steve Lopez tells the story of Nathaniel Ayers, a former student at Juilliard struggling with mental illness, who lives on the streets of Los Angeles playing the violin. Lopez gives Ayers a pen and paper and tells him to write down his thoughts on what it is like to have schizophrenia. The resulting composition exemplifies the challenges that Lopez and his readers might experience trying to understand the logic of Ayers's argument; at the same time the composition achieves its purpose of describing schizophrenia:

> *As a youngster it was very untogether to be labeled mentally ill because of a* [sic] *underlying cigarette habit. That was the root of all evil. I quit and now I feel well enough.*
> *Drug—Abuse*
> *Resistance—Education*
> *D.A.R.E*
> *The treatments for mentally ill persons is from horrible to okay. Overall the idea of deprivation was very effective—making users of drugs realize that there is a real war against drugs being fought every day in LA—LAPD—DARE 911.*
>
> *Los Angeles, California. Recently there was a stabbing incident near the doorway at Lamp 627 San Julian Street. Black man. Mr. Nathaniel McDowell. Los Angeles Times Steve Lopez LAPD 911 Vice Homicide Narcotics. USMC. USN. USA. USAF. USCG. George W. Bush. Command N. Chief of United States Armed Forces.* (80)

Rhetorically, at first glance, Ayers's composition lacks *logos*—there is little to no coherent message and no overtly logical connection between most of the ideas. Ayers relates his mental illness to a cigarette habit and provides various places, names, and abbreviations seemingly in relation to a factual report of a stabbing. It is a challenge for the reader to understand what Ayers means by connecting mental illness with cigarettes and various government agencies.

Understanding Ayers is exactly what Lopez attempts to do in many of his early interactions with him. Lopez struggles in this pursuit because Ayers's life and language have no seemingly inherent logic that he can understand. When Lopez eventually persuades the homeless Ayers to stay at a community housing center a few nights a week, Lopez's first instinct is to find Ayers a psychiatrist, secure a diagnosis for him, and register him for supplemental income. But a doctor Lopez consults cautions against rushing into this logical chain of events, urging Lopez simply to continue being a friend to Ayers. He says, "Relationship is primary. . . . It is possible to cause seemingly biochemical changes through human emotional involvement. You literally have changed his chemistry by being his friend" (210). Eventually Lopez comes to accept that there are "no magic pills" that will cure Ayers and that his friendship may be his most powerful connection to Ayers (284).

Being friends with Ayers causes Lopez to revise his definition of friendship, a revision that hinges less on logic than emotion. He muses, "I don't know if I've truly changed Nathaniel's chemistry. . . . But, yes, of course, we're friends by most definitions of the word. I wouldn't consider talking to him about a career move . . . or the challenges of being a good father and husband. . . . Come to think of it, Nathaniel and I don't really have conversations. Mostly he talks or plays music and I listen. He can't relate to my world and I have trouble relating to his" (211). Still, Lopez maintains that he and Ayers can identify with each other, noting, "We're like each other in many respects," and he compares Ayers's love for music with his own love of writing (211). Despite the ways in which they cannot relate, Lopez reflects, "He's changed my chemistry, too" (211). He seals this realization when Ayers initiates a handshake: "As I leave the apartment one day shortly after he moved in, he calls me back and holds out his hand. It's a long, firm handshake, followed by a smile. I look into his eyes and see the man he's always been behind the racing, spinning madness. The son who lost a father. The musician who lost a chance. No, we don't have too many so-called normal conversations. But what's normal? I hold his hand in mine, and neither of us needs to say a thing" (212). In this handshake, a made-for-Hollywood movie moment, Lopez learns to let go of trying to figure Ayers out and ceases to ascribe a sense of logic to their friendship.[2] He cannot put into words the identification he feels with Ayers and even suggests that a rational explanation based on "normal" standards would not do their relationship justice. He practices a kind of rhetorical listening by trying to relate to Ayers through music and writing, which opens him up to other kinds of identification. Rhetorically, Lopez and Ayers identify with each other and solidify their friendship in the absence of *logos* and instead employ a rhetoric based on touch.

What are the logic and *logos* of touch? Although tactile rhetorical identifications such as those between Lopez and Ayers are not supported in traditional definitions of *logos* based on limited approaches to rationality and reason, other possibilities exist that support a sense of felt *logos*, such as the kind that connects Ayers and Lopez. To explore these possibilities, it is necessary not only to understand the problems that limited understandings of *logos* pose for people with psychological disabilities, but also to recognize the potential of more unstable and generative understandings of *logos* for people with psychological disabilities.

The Logics of *Logos*

The *logos* of an argument—the logical appeals a speaker or writer makes to an audience—is arguably the most powerful part of a message, demanding a rhetor who is rational, reasonable, and psychologically able. As a means of persuasion, *logos* can convince an audience of the truth of a matter, can change minds, and can even physically move audiences to a certain action. *Logos* is, after all, characterized as a drug because of its power in the sophist Gorgias's well-known description: "Speech [*logos*] is a powerful lord, which by means of the finest and most invisible body effects the divinest works: it can stop fear and banish grief and create joy and nurture pity. . . . The effect of speech upon the condition of the soul is comparable to the power of drugs over the nature of bodies. For just as different drugs dispel different secretions from the body, and some bring an end to disease and others to life, so also in the case of speeches, some distress, others delight, some cause fear, others make the hearers bold, and some drug and bewitch the soul with a kind of evil persuasion" (*Ecomium of Helen* 11.8, 14; trans. Kennedy 52–53). As Debra Hawhee, drawing from Jacques Derrida's notion of the *pharmakon,* explains, there is a doubled sense of embodiment at work in this understanding of speech as drugs, in that *logos* both "requires a porous body to pass through in order to incite pleasure or courage, or to induce pain or fear" and "itself bears a body," a body that "moves and mingles with the body/soul it effectively drugs" (*Bodily Arts* 80).

In Gorgias's *logos,* speech, operating as an invisible body or drug coursing through the veins, exhibits explicitly tactile properties, demonstrating the influence of his teacher Empedocles's theory of pores and effluences.[3] *Logos* is a fine-particled entity that circulates through bodies and engraves images on minds. The sense of sight even operates tactilely in Gorgias's *logos,* producing psychological and physical effects, through which the "impressions" of speech "linger" on the body (*Ecomium of Helen* 11.17; trans. Kennedy 54). As Gorgias warns, "It has happened that people, after having seen frightening sights, have also lost presence of mind for the present moment; in this way fear extinguishes and excludes thought. And many have fallen victim to useless labor and dread diseases and hardly curable madness. In this way the sight engraves upon the mind images of things which have been seen" (*Ecomium of Helen* 11.17; trans. Kennedy 54). In his various descriptions of the effects of *logos,* Gorgias argues that *logos* can harm or heal both physically and psychologically. The power of *logos,* as a tactile and visible "body," rests in its potential to drive people to insanity, drug them, and render them completely disabled or to cure, delight, embolden, and enable them. More implicitly, this potential assumes a psychologically able rhetor who can direct audiences to feel and behave one way or another. Who else but a rhetor in control of his or her own psychology could guide audiences to feel such a wide range of disabling and enabling effects?

Aristotle's *logos* also seems to assume a psychologically able rhetor. As George Kennedy explains, for Aristotle, *logos* "often implies logical reasoning" (*A New History* 11), and the most able rhetor employs logical reasoning so well as to render the application of *logos* nearly invisible. For Aristotle, "a statement is persuasive and credible either because it is directly self-evident or because it appears to be proved from other statements that are so" (*Rhetoric* 1356b25–30; trans. Roberts 8–9). *Logos* is "self-evident" in that a rhetor demonstrates it simply by "showing or seeming to show something" (*On Rhetoric* 1356a3; trans. Kennedy 37). In this view, the most logical appeals, and by extension the most logical rhetor, make the best arguments.[4] According to this assumption, a person with a psychological disability would seem categorically excluded from using *logos* in the traditional sense. Associated with rationality, reason, and logic, *logos* seems to possess little flexibility in accommodating rhetors with nonnormative psychologies. Historically *logos* is associated with cultural progress and higher-order cognition. As W. K. C. Guthrie points out, *logos* is often understood as a crucial element in forming the Greeks' "development . . . from a mythopoetic to a rational view of the world" (*Myth and Reason* 5).

Yet, as many scholars have pointed out, an inherent instability of *logos* also exists. As Price notes, Aristotle's "reason" can be interpreted more "flexibly" (*Mad at School* 31). Theorists such as James S. Baumlin and Tita French Baumlin and Susan Jarratt note that the transition from *mythos* to *logos* was not seamless, and in many ways the rationality of *logos* is suffused with the nonrational legends, nonconventional logic, and divine mythologies of *mythos.* As Christopher Johnstone points out in his history of *logos,* reason and rationality are "vexed terms," and to differentiate between a mythopoetic and a rational worldview "is to beg the question, what is 'rationality'?" (28). Rationality, even as it operates in the seemingly fixed category of *logos,* can be relative, situational, and even nonnormative. Yet, as Price demonstrates, explorations of *mythos* that attempt to explore persuasion and *logos* beyond assumptions of rationality and logic often do not adequately support rhetors with mental disabilities. Baumlin and Baumlin, for example, while arguing that *mythos* "reaches beyond conventional logic," also assume that rhetors possess a "health of the soul" that is not readily accessible for people with psychological difference ("Psyche/Logos" 257, 247). As Price describes, in this understanding rhetoric functions as a "healing story," and the "possibility that a mind might be radically *unhealed,* unwell (in conventional terms), that it might be crazy or neuroatypical—this possibility is not offered space even in the expanded version of rhetorical appeals" (*Mad at School* 32).

Despite the tendency to rationalize *logos* and to exclude rhetors with psychological disability, the concept of *logos* possesses fluidity and materiality that potentially broaden its reach, even for rhetors for whom "rationality" or "reasonableness" in the traditional or normative sense may not be possible. Kennedy, exploring the many meanings of *logos* in Greek history, notes that *logos* "is anything that is 'said,' but that can be a word, a sentence, part of a speech or of a written work, or a whole speech. It connotes the content rather than the style . . . it can also mean 'argument,' and 'reason,' and that can be further extended to mean 'order' as perceived in the world" (*A New History* 11). This extension raises the possibility that order can be "perceived" in many ways, perhaps even in potentially nonnormative ways. George Kerferd extends the concept of *logos* further, explaining that there are at least three parts to *logos:* "What we are confronted with is not strictly speaking one word with a number of different meanings, but rather a word with a range of applications," including "an extra-linguistic reference to something which is supposed to be the case in the world around us" (83). He posits this tripartite approach to *logos:* "There are at least three main areas of its application or use, all related by an underlying conceptual unity. First of all the area of language and linguistic formulation, hence speech, discourse, description, statement, arguments (as expressed in words) and so on; secondly the area of thought and mental processes, hence thinking, reasoning, accounting for, explanation . . . thirdly, the area of the world, that *about* which we are able to speak and to think, hence structural principles, formulae, natural laws and so on, provided that in each case they are regarded as actually present in and exhibited in the world-process" (83).

Despite the traditional tendency to rationalize *logos,* a sense of *logos* attentive to its "extra-linguistic" features is more inclusive of the *logos* of rhetors with psychological disabilities such as Ayers. Although Kerferd features ability in his description of the world about which "we are able to speak and to think," this ability need not be exclusively ableist. Extralinguistic features of *logos* potentially include rhetors who do not speak independently and who do not think primarily with reason. These extralinguistic features conjure Gorgias's visceral and tactile approach to *logos,* inviting different ways to listen to and to feel *logos,* yet also invite understandings of *logos* beyond a binary in which it either enables or disables. For example, considering Gorgias's likening of *logos* to drugs, listening rhetorically, with a felt sense of *logos,* to Ayers's seemingly illogical connection between his mental illness and cigarettes and other drugs almost makes sense. Ayers understands the label of schizophrenia as having a physical power over him that will influence the minds of how others see him—if it is "spoken" that he has schizophrenia, that "impression will linger" and have material effects. In relation to Kennedy's interpretation of *logos,* this connection demonstrates a way in which order is perceived in Ayers's world.

For example, when Lopez has an attorney prepare papers for Ayers to sign declaring him schizophrenic and giving his sister control over his finances, Ayers equates being labeled schizophrenic with being lumped together with other "drug addicted thieves," and he verbally and physically threatens Lopez, pushing him down on the floor (258). He is clearly aware of the label's potential power over his life, particularly its power to persuade others that he lacks control over himself and his resources. He declares, "I'm not schizophrenic" (257). It is possible that Ayers rejects this label because he knows the power it holds to define him. To a certain extent he is acting rationally, because he understands the forces at work in his situation. Yet he cannot communicate fully his understanding of the world to others and resorts to physical violence—shoving and pushing Lopez—to communicate. Ayers is neither completely disabled nor enabled by *logos;* instead he uses *logos* both verbally and extralinguistically, to attempt to communicate, in Kerferd's terms, the various "world-processes" or worlds about which he speaks, including those in which he is and is not diagnosed as schizophrenic.

Far from lacking *logos,* Ayers seems to understand himself as caught in the same rhetorical bind as many other people with psychological disabilities. As Anne Wilson and Peter Beresford have explored, patients defined as mentally ill have little to no control over how their words are presented and interpreted, especially in their medical records: "It can feel as if everything you say or do is being taken down and recorded to be used in evidence against you" (148). In resistance to these rhetorical binds, people with psychological disabilities exert agency based in alternative discourses. As Price has shown, people with psychosocial disabilities who write autobiographies often "can and do demonstrate rhetoricability by constructing forms of authority that draw upon, rather than 'overcome,' their disabilities," often refiguring crucial assumptions about autobiographical discourse, including rationality, truth, and independence ("Her Pronouns Wax and Wane" 12–13, 17). One such rhetorical strategy is the use of counterdiagnosis, in which the person with a psychological disability "uses language . . . to subvert the diagnostic urge to 'explain' a diagnostic mind" (17). Ayers exerts a form of counterdiagnosis when he asserts that he is not schizophrenic. From the perspective of Gorgias's definition of *logos,* Ayers seems to possess a sense of rationality and reason because he knows that the diagnosis will change his life.

Furthermore, from the perspective of an extralinguistic approach to *logos,* rhetors with psychological disabilities such as Ayers can and do participate in logical appeals. Ayers's and Lopez's identification through touch functions as a counterdiscourse to the typical readings of *logos* that emphasize a narrow approach to rationality and reason. Lopez initially attempts to figure Ayers out logically but finds more opportunities for identification and understanding through a simple handshake. This handshake operates in all three registers of *logos:* in the area of language; in the area of thought and mental processes; and in the area of the world. Especially as an extralinguistic act, this *logos* of touch conveys a powerful message. Lopez puts the experience in language, but Ayers contributes to the authoring of its message in his initiation of the handshake and by ending it with a smile; he brings it into the area of thought and the world. Ayers brings a semblance of nonrational order and organization to his relationship with Lopez when he offers his hand. The handshake may not be "something that is said" in narrowly verbal terms, but it conveys a message to Lopez of connection and identification.

Beyond Rational: Empedocles's Repetitive Felt *Logos*

Empedocles's approach to *logos* suggests a redefinition of *logos* as a nonrational and extralinguistic appeal based on tactility, ideal for the uses of psychologically diverse rhetors. As explored in the previous chapters, Empedocles's theory of pores and effluences is crucial for understanding rhetoric as tactile. In short, Empedocles's teachings on cosmology and nature posit that bodies and matter—everyone and everything—are the products of four root physical elements: air, earth, fire, and water. The two main energies of Love and Strife, which are interpreted as "psychological" powers by scholars ranging from Guthrie to Simon Trépanier, mix and intermingle the elements and effluences via passages or pores.[5] An endless cycle alternating between Love, which brings the elements together, and Strife, which divides them, continually circulates the elements to create diverse forms of bodies—what Empedocles calls "manifold forms of flesh" (Diels and Kranz 31.B.98; Burnet 219). Although Empedocles's theory of pores and effluences is clearly tactile and is generally accepted as essential to his overall theory of sensation, less explored is the relation of this theory to his rhetoric. As I will explore, Empedocles performs a rendering of *logos* as tactile, an act that models diverse, nonrational forms of thinking and psychology as it explains and values diverse forms of embodiment and cognition.[6] In doing so, Empedocles molds his listeners' minds to be receptive to various forms of *logos* as he teaches his audience to value diverse forms of being, thinking, and feeling. The bodies, souls, and minds that *logos* both requires and bears in Empedocles's teachings are nonnormative and nonrational.

Since Empedocles's general view posits that everything is created by a tactile intermingling among pores and effluences, his sense of *logos* also possesses these fleshy, tactile properties. In one of his most well-known and complete fragments, Empedocles performs a sense of felt *logos* through a repetitive, tactile process:

Double is my account: for at one time it grew to be one alone
from many, at another in turn it grew apart to be many from one.
But double is the coming-into-being of mortals, double their passing away;
for the coming together of all things both begets and kills (life),
and as [all things] grow apart again [life] is nurtured, then disappears (lit. flies away).
And, continuously changing, they never cease,
at one time through Love all of them coming together into one,
at another in turn each carried apart through the hatefulness of Strife.
Thus, on the one hand, in that they have learnt to grow into one from many
And that from the one growing apart in turn they spring,
in this (regard) they come into being and their lifespan is not secure;
on the other hand, in that continuously changing they never cease,
in that (regard) they are forever immobile within the cycle.
But come, listen to my words, for learning increases the thinking-organs.
For just as I said before, speaking the limits (outline?) of my words,
Double is my account: for at one time it grew to be one alone

from many, at another in turn it grew apart to be many from one,
fire and water and earth and the immense height of air
and destructive Strife apart from them, equally balanced in all ways
and Love within them, equal in length and breadth.
Look upon her with your understanding, do not sit with amazement in your eyes,
she who is reckoned even by mortals as born within their joints,
and through whom they think friendly thoughts and perform fitting works,
calling her rightfully (*eponymon*) Joy and Aphrodite.
She no mortal has seen whirling among them (the roots?).
But listen to the deceitless order of the account:
For all of these things are equals and peers as to age,
and each oversees its own honour, and has a particular character,
but they dominate in turn as time revolves.
And besides them nothing more comes-into-being or ceases-to-be;
for if they were continuously destroyed they would no longer be.
And what could increase this totality? And from where would it come?
Or how would they be destroyed into nothing, since nothing is void of them?
But they are (always) themselves, yet running through one another
they become different (things) at different times and are always ever alike.[7]
(31.B.17; Trépanier)

Scholars generally agree that this key fragment, especially the principle that "double is the coming-into-being of mortals, double their passing away," is the "essence" of Empedocles's entire theory (Lambridis 45). Empedocles begins by speaking a "double" *logos*—translated variously as a "double truth" (Lambridis 45), a "twofold tale" (Wright 166), a "double process" (Freeman 53), and a "double tale" (Inwood 223)—and multiplies these *logoi* throughout the rest of the fragment. Each translation, when interpreted in relation to repetitions in the whole fragment, fits the tripartite approach to *logos* described by Kerferd. The arc of the complete fragment also shows Empedocles's overall message and performance of *logos* paralleling the three levels of meaning that Kerferd identifies, comprising the range of applications involving language, mental processes, and the world in general. For Empedocles, these areas of *logos*—language, thinking, and the bodies and matter that make up the world—all intersect tactilely because they are all comprised of pores and effluences. Empedocles's repetitions in the fragment show these three levels of *logos* as interacting physically and tactilely, growing in range of application with each repetition.[8]

Empedocles's repetitive, tripartite approach to *logos* invites nonrationalist and extralinguistic interpretations of *logos*. Exploring Empedocles in relation to nonrational rhetorics, Richard Leo Enos has argued that "theories of rhetoric which pre-date Platonic and Aristotelian rationalism need not, by default, be labeled 'irrational' or 'atheoretical'" ("Aristotle, Empedocles, and the Notion of Rhetoric" 5). Enos describes Empedocles's theories as part of a system of "pre- and non-rational ways of knowing" and argues for understanding Empedocles's influence on the sophists as demonstrated through his "use of style and the epistemology of antithetical thought driving it, the importance of relativism, sense-perception and probability" (19, 5). Particularly interested in antithesis, metaphor, and analogy, Enos posits that Empedocles draws on nonrational and indirect discourse to move listeners beyond "the idealization of a linear, directive process of rationality" (13). For Empedocles, style and epistemology are closely linked; his teachings often perform his theories. Empedocles "taught by example rather than precept" (Guthrie *A History of Greek Philosophy* 135). Functioning as examples, Empedocles's teachings provide "a physical explanation of the universe but also insights into how meaning can be synthesized through the counter-balancing of antithetical notions" (Enos 13). In short, the form and content of Empedocles's use of antithesis work in combination to present his views and to persuade his listeners beyond strictly logical ways of thinking. Repetition functions similarly for Empedocles. In fact Empedocles's repetition often catalyzes audiences to imagine diverse physicalities in the world—especially among human bodies—and to become familiar with diverse thinking styles and psychologies beyond limiting designations of rational, logical, and normal.

As Empedocles repeats his message in his teachings, he teaches his listeners to understand not only his general theory of the cosmos but also how to think about *logos* as a material and tactile entity. In the first repetition of the life cycle of the cosmos, he wants his listeners—his pupils—to be able to hold two somewhat conflicting principles in tension. Signaling this tension, he relates, "Double is my account" and says, "on the one hand," that everything is connected, in that all beings and matter "have learnt to grow into one from many." Simultaneously he wants his audience to understand a competing principle: "on the other hand," everything is "continuously changing." These two seemingly antithetical principles of sameness and difference form the foundation of Empedocles's worldview and teach his listeners to think beyond simple logics and binaries. He repeats his argument different ways and multiple times not only to teach his audience a different principle but also to train their minds to be able to hold the logic of these competing notions together at once. He continues, reiterating the doubleness of his *logos,* "But come, listen to my words, for learning increases the thinking-organs. / For just as I said before, speaking to the limits . . . of my words, / Double is my account." After elaborating on the principles of sameness and difference, he adds new information into the mix in this repetition, including the images of "fire and water and earth and the immense height of air" and the powers of "destructive Strife apart from them" and "Love within them." In this repetition Empedocles adds the four elements and the two psychological forces of Strife and Love, challenging his audience not only to hold in tension the competing demands of his first principle of sameness and difference but also to understand the function of the additional elements and psychological forces in his larger cosmological theory.

In effect, Empedocles's repetitions are not simply repetitions of the same *logos* but are also accretions of double, triple, and even quadruple *logoi* that add new information and concepts. Accordingly, Empedocles's *logos* functions pedagogically through form as much as content—Empedocles guides his audience toward how to think, not just what to think. The content of his teachings in this crucial fragment shows that *logos* is multiple, diverse, and even contradictory. Empedocles's repetitions of this lesson also model for his audiences how to receive this argument, including how to think through and feel *logos* as a dynamic, nonrational, and flexible concept. He attempts to mold his audience's minds—their "thinking-organs"—with his words, using repeated learning to increase wisdom. This repetition primes the bodies and the minds of his listeners to register *logos* as a combined experience of thinking and feeling.

Thought is a felt process in Empedocles's theories, shaped by both difference and repetition. In other fragments Empedocles describes cognition as always embodied and shaped by a repetitive process. He identifies the heart as the thinking organ, "dwelling in the sea of blood that runs in opposite directions, where chiefly is what men call thought; for the blood round the heart is the thought of men" (31.B.105; 220). A process of habituation that includes repetition shapes thought through the body, resulting in difference. As Empedocles describes, "wisdom in men grows according to what is before them" (31.B.106; 220), and "for out of these are all things formed and fitted together, and by these do men think and feel pleasure and pain" (31.B.107; 220). Thinking and feeling are "fitted together" similarly to how pores and effluences combine in Empedocles's overall theory, with this tactile "fitting" resulting in differences in psychologies and cognitions for his listeners: "And just so far as they grow to be different, so far do different thoughts ever present themselves to their minds" (31.B.108; 221). Difference is crucial to this cognitive development; as M. R. Wright describes, a "satisfactory mixture of bodily elements" is "conducive to thought, which thrives in the appropriate environment" (236). In other words, regarding Empedocles's pupils, "what is before men"—their various environments, the concepts they learn, and different experiences—changes their thoughts and thought processes. This process also involves reciprocity: "the external condition affects the growth of the thinking, and . . . an internal change of structure results in a change of thought" (Wright 236). In Empedocles's additional remarks on the heart, repetition functions crucially in the process of "a sifting of the *logos* in and around the heart" by which "the thoughts thus received then increase and strengthen" (Wright 164). Empedocles's sense of *logos*, felt in the heart as much as exhibited by one's cognition, is physical, psychological, and embodied. Repetition and habit, molded by difference, function to shape *logos* in concert with the body and the mind that receive *logos*.

Outlining a pedagogical practice for his own teachings, Empedocles reinforces this process of repetition and habit through an attention to tactility. Counseling his audience on how to receive his teachings, he advises, "If you push them (*eresias*) firmly under your crowded thoughts (*prapidessin*), and contemplate (*meletēisin*) them favorably with unsullied and constant attention, assuredly all these will be with you through life, and you will gain much else from them, for of themselves they will cause each thing to grow into the character (*auta gar auxei taut' eis ēthos hkaston*) according to the nature (*phusis*) of each" (31.B.110; Wright 258; Hawhee "Bodily Pedagogies" 150). Empedocles uses tactile and embodied language to advise his listeners, urging them to "push" his teachings "firmly" under other distracting thoughts that may crowd their minds. The verb *eresias* connotes the effects of touch, including the "force of 'push,' 'thrust,' and . . . 'struggle'" (Hawhee 150). As Hawhee argues, for Empedocles, the methods of study for rhetoric are a bodily production, or "a mutually constitutive struggle among bodies and surrounding forces" (150).

This bodily production is closely linked to the senses—especially to touch—and extends to both cognitive and psychological productions, encompassing a whole-body approach to learning that acknowledges the partiality of experience. Describing the psychological force of love, Empedocles advises his pupils, "Look upon her with your understanding, do not sit with amazement in your eyes," a reinforcement of a similar admonition for his listeners, in which he also cautioned them to keep in mind the partiality of their senses and understanding: "For straitened are the powers that are spread over their body parts, and many are the woes that burst in on them and blunt the edge of their careful thoughts! They behold but a brief span of a life that is no life, and, doomed to swift death, are borne up and fly off like smoke. Each is convinced of that alone which he had chanced upon as he is hurried every way, and idly boasts that he has found the whole. So hardly can these things be seen by the eyes or heard by the ears of men, so hardly grasped by their mind! Howbeit, thou, since thou has found thy way hither, shalt learn no more than moral mind hath power" (31.B.2; 204).

Using an approach based on touch, Empedocles urges his listeners to use all of their potential resources for learning, including sensory perceptions, to increase the powers of their minds. Emphasizing listeners' "powers that are spread over their body parts," he encourages a whole-body approach to cognition based on touch, with the caveat that this approach is necessarily partial. For Empedocles, "the 'devices' for understanding . . . are the sense organs, with that of touch being spread over the whole body" (Wright 156). Touch is spread over the entire body because each sense, in Empedocles's theory, operates via touch. Sight, for example, occurs when "the round pupil, confined within membranes and delicate tissues," is "pierced through and through with wondrous passages" of fire that make images (31.B.84; 217). Hearing operates through the "fleshy shoot" of the ear, which when set in motion, "drives the air against the solid parts and makes an echo" (A.86.9; Inwood 197).

Empedocles also forwards a whole-body approach based on tactility when he counsels his listeners not to "speak recklessly . . . then sit on the high throne of wisdom" but to "observe with every power in what way each thing is clear, without holding any seeing as more reliable compared with hearing, nor echoing ear above piercings of the tongue; and do not keep back trust at all from the other parts of the body by which there is a channel for understanding, but understand each thing in the way in which it is clear" (31.B.3; Wright 160). Drawing his audience's attention to both the power and the limits of their bodies, Empedocles explains that sensory perceptions through the eyes and ears are useful but that his listeners must also learn that these perceptions are limited and only parts of the whole. In fragments such as these, Empedocles employs tactile imagery and draws on his generalized theory of the mixture of pores and effluences to encourage audiences to explore all their opportunities for understanding. Empedocles encourages audiences to keep their minds and bodies open to all channels of experience, reinforcing his identification of the channels of the pores through which effluences move, and warns audiences against "blunting" the edge of their thoughts or relying on single senses that may block other channels of understanding.

In his next repetition of his main message regarding his cosmological theory, Empedocles builds on his tactile approach by repeating previous information, but he adds a sense of movement to the four physical elements and two psychological forces, challenging his audience to think more flexibly with more sophisticated concepts and preparing them to understand the diverse forms of bodies and minds he describes to them. In these lines Empedocles adds even more principles to the mix for his listeners, once again modeling a thought process whereby his listeners must hold even more *logoi* in tension.[9] He repeats,

> But in Love we come together (to form) a single cosmos
> and in Strife in turn it grows apart to be many from one,
> from which (many) is everything, as many as ever were, as are, and as will be hereafter;
> and (from which) trees have sprung, and men and women,
> and beasts and birds and fish which thrive in water,
> and long-lived gods mightiest in honours.

> But under her (Strife) they never cease leaping about continuously,
> in close-packed whirls . . .
> without pause . . .
> . . .
> but leaping about in all directions they never cease
> . . .
> but changing continuously they leap in all directions within a circle.
> . . .
> In just this way all things ran through one another. (Martin and Primavesi; Trépanier)

By adding a sense of movement to the elements and forces, Empedocles both builds on his earlier teachings and extends them, challenging his listeners again to think as flexibly as the contents of his teachings demand. Again he uses a *logos* that is tactile in nature, with various elements "leaping" and springing and "close-packed" whirls and effluences running through one another via the opposing but complementary psychologies of Love and Strife. Empedocles animates his *logos* with this description, challenging his listeners to keep track of many different moving parts and guiding them to hold these thoughts in tension.

In a subsequent repetition, Empedocles most clearly makes his performance of a tactile *logos* apparent to his audience and models the flexible kind of thought he expects from them. Using tactile vocabulary, he elaborates on the "close-packed whirls" and notes that from this mingling "things close-pressed and solid" issue forth, separated by Strife and brought together by Love (31.B.21; 209). He declares, "For there are these alone; but, running through one another, they take different shapes—so much does mixture change them" (31.B.21; 209). A physical and tactile sense of movement directs this running, mingling, and mixing of change and difference. He repeats his principle of sameness and difference, saying that "all of these—sun, earth, sky, and sea—are at one with all their parts," but he adds new information by describing the process by which certain parts are "more adapted for mixture" than others whose "forms imprinted on each" are more hostile (31.B.22; 209).

Empedocles elaborates on this process in another fragment with a tactile metaphor involving painting: "Just as when painters are elaborating temple-offerings, men whom wisdom hath well taught their art,—they, when they have taken pigments of many colors with their hands, mix them in due proportion, more of some and less of others, and from them produce shapes like unto all things, making trees and men and women, beasts and birds and fishes" (31.B.23; 209–10). This image of painters mixing colors with their hands in differing proportions is crucial to the type of thinking that Empedocles models for his listeners as he transitions to the ultimate goal of his cycle of repetition. He is preparing his listeners—whom he has taught to be able to hold multiple *logoi* in tension, particularly through repeating the principle of sameness and difference while adding elements of additional forces and movements—also to be able to understand that different bodies, both able and disabled, proliferate as widely as the forms of thinking that he models for his audience. As Empedocles's *logoi* proliferate, so do the types of bodies and minds he adds to his worldview, which in turn are matched by the expanded embodied mindfulness that his audience cultivates.

Empedocles's ultimate goal in his cycle of repetition is to teach his audience to think and feel in ways that are flexible and diverse so that they are prepared to accept and value the wide range of bodies and cognitions that proliferate in the cosmos. Empedocles's evoking of *logos* as a sense of proportion illuminates this objective. As Guthrie points out, *logos* can mean "proportion" in Greek, a valence of the term that, considered in connection with Empedocles's teachings, such as his example of the painters, solidifies *logos* as material and tactile. Guthrie writes, "This proportion is a chance outcome of the interaction of Love and Strife. Now the word for proportion is *logos,* but *logos* had many other meanings in Greek, some of which have reference only to the behaviour of rational beings: it can mean thought or the result of thought" (*A History of Greek Philosophy* 160). Despite Guthrie's emphasis on rational beings, Empedocles's evoking of *logos* as proportion in his final repetitions suggests extensions beyond the rational.

To exceed rationality, Empedocles again draws attention to a sense of proportion, as he did with the manual actions of the painters and their colors, to model how love and strife circulate to produce a diversity of bodily and mental forms, all intimately connected. In a culminating repetition of the cosmic cycle, Empedocles models ways of thinking and being that exceed reason and rationality. In this repetition—"now I shall retrace my steps over the paths of song that I have traveled before, drawing from my saying a new saying"—Empedocles details that "many things remained unmixed" in this part of the cosmic cycle, "alternating with things that were being mixed," such that "some . . . still remained within, and some had passed out from the limbs of the All. But in proportion as it kept rushing out, a soft, immortal stream of blameless Love kept running in, and straightaway . . . those things were mixed that had before been un-mixed. . . . And, as they mingled, countless tribes of mortal creatures were scattered abroad endowed with all manner of forms, a wonder to behold" (31.B.35, 36; 211–12).

The movement and activity of the mixing elements "in proportion" to each other prepare listeners for Empedocles's descriptions of the formation of bodies—"all manner of forms"—which exceed the confines of normality or rationality. From the earth, he says, "many heads sprung up without necks and arms wandered bare and bereft of shoulders. Eyes strayed up and down in want of foreheads" (31.B.57; 214). In addition, "solitary limbs wandered seeking for union," and "shambling creatures with countless hands" (31.B.58, 60; 214) issued forth. Furthermore, "many creatures with faces and breasts looking in different directions were born; some, offspring of oxen with the faces of men, while others, again, arose as offspring of men with the heads of oxen, and creatures in whom the nature of women and men was mingled, furnished with sterile parts" (31.B.61; 214).[10] In this final repetition, Empedocles at last brings his various principles to bear on the production of bodies, which take on a wide variety of forms. Diversely shaped bodies form as Empedocles's *logoi* proliferate.

Empedocles also builds psychological difference into his modeling of *logos* as proportion, particularly in his theories of the formation of bodies, which include attention to indirect, nonrational, and nonlinear ways of thinking. By featuring the god Hephaestus, who was widely known as the "crippled god," in the process of making bones and blood, Empedocles works the dexterity of a nonnormative body and mind into his evoking of *logos* as proportion. He explains, "The kindly earth received in its broad funnels two parts gleaming Nestis out of the eight, and four of Hephaistos. So arose white bones divinely fitted together by the cement of proportion" (31.B.96; 218). The "*logos* of bone" is four parts fire, two parts earth, and one part air and water, with the excess of fire accounting for the dryness and whiteness of bones (Wright 209–10).[11] As Marcel Detienne and Jean-Pierre Vernant have shown, Hephaestus, with his curved hands and feet, is a model for "cunning intelligence" in Greek society, a kind of thinking that is not straightforward, linear, or strictly rational. "The particular shape of his feet is the visible symbol of his *mētis,* his wise thoughts and his craftsman's intelligence," which are demonstrated in qualities such as "pliability and polymorphism" and signify that which is "pliable and twisted . . . oblique and ambiguous as opposed to what is straight, direct, rigid and unequivocal" (Detienne and Vernant 272, 46). Extending the significance of Hephaestus's way of thinking and moving for the contexts of rhetorical history and theory, Jay Dolmage argues for understanding Hephaestus as a model of disability rhetoric, rereading the stories of Hephaestus as a "challenge to stories of rhetorical history that reinscribe normative ideas about rhetorical facility and about which bodies matter" ("Breathe Upon Us an Even Flame" 119). Hephaestus provides "an image of disability as valued by ancient society" and "recover[s] a different rhetorical body" for contemporary audiences, one that is not "composed in perfect ratio" (119, 125). In the context of Empedocles's teachings, Hephaestus's body can also be seen to recover a different kind of mind and psychology for rhetoric. Empedocles's evoking of Hephaestus in the making of bones suggests a way to make a connection between Hephaestus's disabled body and his cognition and psychology. Since Empedocles's teachings rely on the intimate connections between thinking and feeling in his naming of the heart as the seat of thought—"the blood round the heart is the thought of men" (31.B.105; 220)—it follows that bones fashioned by the curved hands of Hephaestus would support different types of thinking styles, cognitions, and psychologies. Bones fitted together by the differently formed but dexterous hands of Hephaestus illustrate proportions of *logos* and psychologies beyond normality or rationality—circuitous, recursive, and unpredictable.

The goddess of love Aphrodite works with Hephaestus to perform this tactile and manual *logos* construction in the making of blood and flesh via specific proportions. As Empedocles describes, "The earth, anchoring in the perfect harbors of Aphrodite, meets with these in nearly equal proportions, with Hephaistos and Water and gleaming Air—either a little more of it, or less of them and more of it. From these did blood arise and the manifold forms of flesh" (31.B.98; 218–19). As Wright describes, the "physical constitution of thought" also includes the four elements of fire, earth, water, and air in "almost equal proportion" (209). Since Empedocles connects blood and thought so closely, the arising of "manifold forms of flesh" signals a similar proliferation of manifold forms of thought, thinking, and psychology. Hephaestus, together with Aphrodite, shapes the proportions of the elements, forming a fleshy *logos* of blood, bone, and thought. Aphrodite coordinates the fitting of these proportions, from blood to thought, in a tactile manner. In other comparable fragments Aphrodite is similarly manually active, "fitting together," "nailing," "gluing," "molding," "working with her hands," and "generally being busy" (Wright 238).[12] Aphrodite's and Hephaestus's hands, nondisabled and disabled respectively, form the proportions from which bodies and minds are made, modeling diverse embodiments and psychologies.

In theory and practice, Empedocles persuades his audience that the content and function of his teachings are to explain and help shape a wide range of diverse bodily forms, psychologies, and types of cognition. With his successive repetitions and elaborations of his theory of the cosmos, Empedocles has already molded the minds of his listeners significantly, preparing them to hold multiple and competing *logoi* in tension. These proliferating forms of thought that he models and inspires for his audience prepare them to accept the wide range of bodies, cognitions, and psychologies that are possible. By repeating his *logoi,* Empedocles teaches his listeners to think beyond rationality and in flexible ways, preparing them to understand how diverse bodies and psychologies constitute the human spectrum. This sense of understanding entails both thinking and feeling, as Empedocles models a sense of felt *logos* for his audiences, teaching them to be attuned to the embodied processes of learning and communicating.

Interfacing and Rhetoricizing Touch Online

Empedocles's pedagogical style invites present-day audiences to enact a type of rhetorical listening similar to that defined by Ratcliffe and suggested by rhetoricians such as Lewiecki-Wilson, Prendergast, and Price for the specific needs of people with mental or psychological disability. Empedocles's *logos,* operating beyond rationalism and in extralinguistic registers, provides models for valuing acts of rhetorical identification between people with psychological disabilities and their audiences based on the sense of touch, such as the handshake between Ayers and Lopez. Empedocles's theories illuminate the ways in which people with psychological disabilities—long considered to be lacking *logos* or appeals to rationality—exhibit a method of felt *logos* in their rhetorical efforts, in which they draw attention to the function of *logos* as an interface for reality.

Online support spaces for people living with psychological disabilities are sites of rhetorical listening in which people use a technological interface to negotiate both the physical and psychical challenges of psychological difference and to form identifications with various audiences. As Prendergast suggests, it is important to listen to the "unexceptional schizophrenic," which can be interpreted to include people with schizophrenia who write on online forums. Extending Lewiecki-Wilson's suggestion to listen rhetorically to people with psychological or mental disabilities, it is possible to position online forums as places where people with disabilities listen rhetorically to each other. In their negotiation of sensory perceptions, including the challenges of putting words to them, people with psychological disabilities—from schizophrenia to depression—turn to the sense of touch in real-life and virtual spaces to identify with others and to shape and reshape *logoi* in the broadest sense of the term—as words, the world, and mental constructs.

Empedocles's theory of pores and effluences is based on a notion of the interface that is particularly relevant to online spaces such as psychological support forums. Siegfried Zielinski, in his "deep time" archaeological study of media, suggests that Empedocles's ancient theories of perception "have some bearing upon the frenetic contemporary sphere of activity that is theory and praxis of media: the interface between one and the other, which can be defined as the interface between media people and media machines" (53). Applying Empedocles's theory to new media interfaces, Zielinski writes, "Interpreted technologically, it is a theory of double compatibility: size and relative power of the pores and effluents must match so that exchange can take place. Physically, it is a theory of affinities, which can be described in psychological terms as a concept of reciprocal giving and receiving of attention" (53, 55). Combining these physical and psychological interfaces results in a model of technology based on reciprocity. The reciprocal act of giving and receiving structures both physical and psychological processes, inviting applications to modern media interfaces in which *logos* is fluid, nonrational, and driven more by feeling and identification than by reason and rationality.

Empedocles uses the forces of Love and Strife to draw attention to the crosscurrents of psychological energy that direct the physical elements of world and word building, establishing this interface of *logos* as necessarily a tactile site in which people, words, and things interact. Love and Strife are psychological forces that model a sense of give and take, or reception and direction of attention, and that shape Empedocles's construction of *logoi* as much as bodies do. For people with psychological disabilities, many of whom struggle to understand the difference between what is real and what is unreal, the interface between people, words, and things is an especially crucial site for defining oneself and forming identifications with others. Online support spaces for people living with psychological disabilities function as interfaces in which people mediate both the physical and psychical challenges of psychological difference. In their writing and their support of each other, people with psychological disabilities call attention to the sense of touch as a way of making sense of their worlds and rhetoricize touch in order to respond to different social contexts, connect with their audiences, and establish themselves as possessing an alternative sense of felt *logos* in both their virtual and real lives.

Although psychiatrists consider tactile hallucinations somewhat rare, especially in comparison with other types of visual or auditory hallucinations, in online spaces people with schizophrenia frequently share their experiences with tactile hallucinations, requesting support from others and demonstrating how they make decisions about the difference between reality and hallucinations. In one online mental health forum, a writer posted a message titled "I Think I Might Have Schizophrenia" and described a tactile hallucination. The poster, Emily, wrote, "i told my dad about the voises [*sic*] and the bugs crawling in my skin and he gave me a look like there was something wrong and told me it was all in my head and i would like to know if it is all in my head or if there yreally [*sic*] is something wrong" ("I Think"). Several respondents replied with possible diagnoses and recommendations to seek medical advice, while others emphasized the importance of "listening" to oneself and finding a trusted adult with whom to speak.

In a response two days after her original post, Emily reiterated, "My parents wouldn't believe me :cry:" ("I Think"). One concerned respondent in particular offered concrete advice, advising her to "keep in mind that schizophrenia is very hard to diagnose, because hearing voices and seeing things and other psychotic symptoms can be part of other conditions as well: depression, bipolar disorder, etc. If your parents are not going to take you seriously, then you need to talk to a teacher or the school counselor. Are you in school? I recommend printing out this post you wrote here and showing it to the school counselor or a teacher you trust. From that point, they will be able to have a serious talk with you and your parents and help them understand the seriousness of all this and that you should be seen by a psychiatrist" ("I Think"). Rather than telling Emily that her hallucinations could not be real or that she is unbelievable—essentially charges that she lacks *logos*—this commenter worked with what Emily wrote, whether it was traditionally logical or not, encouraging her to print out her post to the online forum and show it to someone she trusts. In effect, this suggestion, which many others echoed on the forum as sound advice, operates as a way of validating Emily's sense of *logos* and encouraging her to explore it more, gaining feedback and a sense of trust with a teacher or counselor. In the continuing thread, three respondents, who each revealed that they have been diagnosed with schizophrenia, identified with Emily, and others spoke from the experience of having family members with schizophrenia or bipolar depression; all of them encouraged Emily to explore the full range of what her hallucinations could mean. A wide range of interpretations of her symptoms was offered among these respondents—depression, bipolar disorder, hormone imbalances, and kidney and liver problems are all mentioned as possible causes of her symptoms in addition to schizophrenia.

Despite the range of opinions, a common thread among posters is that Emily needs to continue to talk about her symptoms and not simply write off her experiences, especially the tactile hallucinations, as crazy. One respondent, "speaking from experience," urged Emily, "If you are at all worried about anything in your life that could manifest itself in anyone [*sic*] of these conditions then use this discussion board to vent it, or approach someone you trust, it doesn't have to be a doctor. . . . We are all hear [*sic*] to listen to anything you have to say, to help you without judging you. . . . Please come back on line and talk to us!" ("I Think"). Another poster reinforced this sense of the forum as a safe space for identification, writing, "I know I'm not crazy & you have to know that too; they're just attacks & deceptions . . . I've been through years of torment; you can get help now. Please let us know how ur [*sic*] doing; we are here 4 you. U have to know you're not alone" ("I Think"). Respondents, often identifying with Emily, encouraged her to continue to write about her experiences both online and in her real life, calling attention to and blurring the boundaries between real and virtual interfaces. In particular, the advice to print out and share her account of her tactile hallucinations and to continue to participate by writing in the forum operates as a rhetoricization of the sense of touch in both real and virtual spaces. In Empedoclean fashion, style and content merge in this approach—posters encouraged Emily to repeat her experiences with her hallucinations so that she can have multiple opportunities to sift through the meanings and implications of those experiences. By printing out her post and handing it to a counselor or teacher, for example, Emily's *logos* is not only validated but also duplicated in additional contexts and rhetorical situations. The discussion of touch online and in real life shows respondents rhetoricizing touch according to purpose, context, and audience, all while forming identifications among each other.

On other mental health forums, writers with schizophrenia work together to discuss and evaluate their tactile hallucinations, often rhetoricizing the sense of touch by discussing it, using it to create identifications with others, and then collaboratively sorting through reality with it. On one online support group for schizophrenia, for example, a poster named ConstantKnot wrote, "So i've been told to post on here. . . . But i have to disagree with why i should write here, I am not ill, I am not schizophrenic, I am not mad. . . . I don't really know what to write here, But i suppose I could tell you about my life, But they wouldn't like it" ("I Told You"). He continued, describing motivations including desiring to "slaughter the world" and "torture the weak," but attributed these urges to external forces:

> I think i'm okay. But then is this just the reality which the suits have created for me? Because they have created a good reality for the "norms." Does the "unnorms" get a bad reality, Until you see the glitch. You see the real reality. . . . The world is burning but the atoms from the ashes, have been programmed to react with our brain waves because of bugs under our skin. The suits have plannted [*sic*] these bugs. I managed to cut mine out. So i get glitches because there is still one left under my skin, on my back where i can't reach it. But the suits know that two of the three bugs have been destroyed, so there [*sic*] following me. . . . They have tried to drug me down with there [*sic*] poisons. But I understand that these poisons they try to give me, will take away all my thoughts. They try to tie you down with these poisons so you cannot think, so you cannot save the world. . . . This is how life will end. This is how the world falls to the ground. The only person who can stop it. Is writing this. ("I Told You")

From the traditional perspective of *logos* as reason and rationality, it is tempting to evaluate this post as *logo*-less, paranoid, or simply crazy. The original poster believes that the suits have implanted bugs under his skin and have changed reality. Like many people with schizophrenia, including Ayers, he is paranoid and believes that he is being drugged, poisoned, and followed.

From the perspective of felt *logos,* however, the poster was clearly using tactile hallucinations to negotiate the difference between what is real and what is unreal, with several accurate assignations. Astutely he asked critical questions about the differences in reality between "norms" and "unnorms." Theorists and activists in disability studies have asked similar questions in the last three decades, especially in their focus on how certain labels or diagnoses result in real material differences in the realities of people with disabilities. More important, however, the poster believes in the importance of writing theories and thoughts down, sharing his with other support group members. One respondent connected particularly with the original poster's search to sort through reality: "The reality you are in writing this, is not the true reality and there truly are no bugs" ("I Told You"). Although the original poster had expressed reluctance to write about his experience and was told by another poster that his reality is not "true," he received affirmation that recording his struggles with reality, including his tactile hallucinations, leads to opening his thoughts up to review and interpretation by others, many of whom have had experience with his condition. Most productively this demonstrates that writing about tactile hallucinations can be an effective way of interfacing one's thoughts with others.

Similarly, on a thread on schizophrenia.com, many users discuss their experiences with tactile hallucinations, especially those with bugs, and collaborate with each other in questioning and even ascertaining reality. On schizophrenia.com's support forum, one user wrote, "Has anyone had the sensation of bugs in their brain, moving around nesting?" and received ten responses, each echoing the others and surprised to find others with whom to identify ("Delusions"). One user wrote, "Having read this thread, I know now I'm not the only one. I told my psychologist I had bugs under the skin and he said only drug addicts have this" ("Delusions"). Another replied, "I've had alot of tactile hallucinations. . . . All i could tell you is to keep reminding yourself that its [*sic*] not real. Its [*sic*] just not real" ("Delusions"). Respondents to Emily's, ConstantKnot's, and others' posts about tactile hallucinations do an important service by forming identifications and solidifying the forum as a space for frank discussion about psychological disabilities and differences, especially when medical professionals do not listen to their symptoms. Posters and respondents on psychological support forums work together to feel out a sense of the *logos* they inhabit. By putting words to their tactile experiences and sharing them with others, they rhetoricize their experiences with touch, placing them in contexts with supportive audiences, forming identifications with others, and engaging in productive discussions of reality.

The active writing and collaboration strategies shared on schizophrenia forums also occur on support forums for people with depression and also feature the sense of touch. One poster, named Dez, began a thread on a depression and social anxiety forum by asking, "How many of you are touch starved? This is something that has been on my mind for a long time. I'm not saying it's a miracle cure but can touch have a healing effect for sadness? And can lack of touch in someone's life contribute to their depression?" ("Lack of Touch"). Welcoming contributions from other people with depression, he identified himself as "extremely touch starved" yet unable to deal with accidental touches such as on the bus when sitting next to someone or when someone brushes an arm against him, calling it "almost alien" to him. He asked, "Does anyone else have a similar problem? How do you deal with it?" These questions generated thirty responses on the forum, with various repetitions of the theme of touch in depression.

Almost every respondent identified with Dez's hunger for touch and yet, like Dez, also struggled with how to manage touch. One wrote, "This is totally a major issue for me. For whatever reason, I am afraid of touch, yet I want it so bad. I would even say I need it to move forward with my life" ("Lack of Touch'). Another wrote, "Yes, yes, yes, yes, YES! What a great topic. It describes me exactly. I am always uncomfortable with touching and do feel 'touch-starved'" ("Lack of Touch"). Many other respondents identified with this issue and Dez, even repeating the original language of his initial post: one wrote, "I couldn't have said it better myself," while another agreed, "your original post . . . describes me exactly, almost as if I had written it" ("Lack of Touch"). Together respondents sorted out their complicated feelings regarding touch, finding correlations between touch and their depression. One wrote, "I do feel 'touch-starved' and that's such an apt expression. The lack of touch undoubtedly leads to further alienation. It's almost like a cycle. We don't touch people because we're shy and socially backward but then over time the fact that we are so touch-starved only serves to increase our shyness and leads to depression" ("Lack of Touch"). One respondent replied, "I'm in the exact same situation as you," and another drew comparisons between physical and emotional support, stating that "human connection, physical or verbal," is "strengthen[ing]" ("Lack of Touch"). After voicing their identification with Dez, other posters transitioned to suggesting alternative and potential solutions to address the problem of touch in their lives. One suggested that other posters write down their experiences with touch daily:

> Do you know why babies need touch? Think about it. It tells them that they are not alone. . . . A human is a human is a human; we are just as human as a tiny newborn baby is. Like it or not, we depend on other people. We are social "animals" just like, say, wolves are. We were born to be in a "pack" and live closely with others. That is why we are in so much pain, why we are so lonely. We are meant to be touched and loved and held and we are meant to live in a family—not alone! . . . Now that we are adults, and supposedly free, we wonder why are we so afraid of asking for a hug—from our own sister or friend, or of giving a pat on the shoulder—to someone we greatly admire or feel appreciation for? It's because it was learned. How to unlearn it? Make a log this week of how many times you want to touch someone or were touched, your feelings at that moment (rate your feeling's intensity 1–100%), and all of your thoughts that occur in your mind at that moment. At the end of the week, look at your thoughts and question them. Ask yourself, How true is that thought? This way, it will be revealed that it is your thoughts, or rather your irrational beliefs, that cause you pain. In time, you can catch your thoughts and change them more easily to thoughts that are more closer [*sic*] to reality. ("Lack of Touch")

Although this poster credited his counselor with the touch log assignment, in his post he adjusted the touch log for his audience of fellow online support group writers and contextualized it to encourage participation. In addition he positioned the writing of the touch log as a way to "catch" one's thoughts, a process that depends on physically writing down experiences with touch so as to make them a more concrete reality. In doing so, he hoped to stem "irrational beliefs" and foster thinking more attuned to "truth" and "reality." In Empedoclean fashion, this repetitive exercise is a mindful and embodied practice by which posters change their cognition by increasing their wisdom according to "what is before them" (31.B.106; 220). Together they sift through different *logoi,* expanding their ranges of thinking and feeling through reflection and repetition. The poster encouraged others with psychological differences and disabilities to rhetoricize touch by writing about instances of touch in their lives, reflecting on their thoughts, and evaluating their feelings. They create the online space as an interface for processing touch, thought, and feeling in words, a rhetorical action that carries over into their real lives.

In ways such as these, people with schizophrenia or depression use online interfaces to approximate the kinds of connections that Ayers and Lopez make in real life. They connect to each other, listen to each other rhetorically, and feel out the implications of a competing sense of *logoi* in their lives. The felt sense of *logos* they cultivate by discussing touch online also resonates in their offline lives. In fact the efforts that online users with psychological disabilities make to connect with each other are similar to the unconventional but effective therapies that psychologists such as Lauren Slater use in residential communities of people with schizophrenia. Forgoing conventional skills-based approaches to therapy, Slater depends on repetitive instances of touch to build connections among patients and to facilitate opportunities for identification among them. In one group meeting, she took seriously a patient's claim that a spaceship had landed on his stomach and encouraged all the patients to touch his stomach, taking a ride in the spaceship, as she asked them what they saw and how they felt. She encouraged another patient who was feeling lonely to talk to and hug each of his friends.

After these exercises, a particularly unreachable patient named Moxie, who had engaged in self-injury, revealed his scars, touching each one in front of the other patients while inviting them to touch his scars as well, then shaking each patient's hand and repeating, "welcome to my country" with each encounter (Slater 35). Showing his scars and reaching out to his fellow residents, Moxie initiated others into his world while attempting to reach into their worlds as well. Slater concludes, "Crazy or sane, we all know the desire for skin touching skin," and she believes that "simple touches" such as the ones the patients engage in mean that "their stretches for connections are occurring at more complex levels, are rooted in the need to belong, to participate in the wider world" (17–18, 19). Repeated acts of touch—from handshakes to hugs—bind the patients to each other and the community. In Empedoclean fashion, different *logoi* proliferate as each of these different touches occurs, demonstrating to Slater and the patients the wide ranges of meanings available for them to spur identification and connection. The patients Slater encounters do not suddenly exhibit rationality or reason in the traditional sense of *logos,* but touch and its repetition operate as vehicles for attending to a different kind of felt *logos* that restructures relationships at the residence program through connection and partial identification. Similar to contributors on online psychological support forums, the patients at the residential home sort through their lives and *logoi* by feeling their way through conflict and by establishing connections with each other.

Contributors to online psychological support forums and residential units such as these are models for a sense of the feel of *logos,* as they negotiate multiple and even contradictory experiences that do not fit neatly into rationalist paradigms. Like Empedocles's proliferating and tactile *logoi,* people with psychological disabilities writing on online forums and communicating in residential units construct proliferating responses to the sense of touch in their real and virtual lives, proportioning out reactions and responses to touch that are productive and lead to identifications. Although there is a desire to sort through reality and truth, they typically eschew either-or binaries of rationality in favor of exploring what their encounters with nonrational, irrational, or simply contradictory sensory perceptions can mean. Rather than silencing their tactile hallucinations or their contradictory responses to touch, they share them with others online and in real life, generating multiple responses and possibilities for interpreting their worlds. In this way they model alternatives to the traditional understanding of *logos* as a rational or logical appeal by working with the complexities of *logos.* Instead of creating a binary sense of logic, people with psychological disabilities use touch to create a variety of different interpretations, feeling their way toward a sense of *logos.* Contributors on online sites in particular collaborate not only to support each other but also to maintain the online support interface as a viable medium for influencing both virtual and real life. In Empedoclean fashion, they call attention to the physicality of their worlds and words, both in real and virtual life. Touch, as it functions as a literal sensation in real life or how it is rhetoricized online, is the interface that connects people and builds identification. This identification binds participants, connecting them to each other and contributing to their appreciation of different psychologies.

4

Habituating *Ethos*

Touch, Autism, and *Mētis*

People with autism, frequently diagnosed as possessing "deficits" in social interaction and communication and stereotyped as "asocial" or "alone in their own worlds," present particular challenges to the limits of traditional rhetorical models of identification. As Jim Sinclair explains, "Autistic people are generally seen as lacking in ability to share common interests with others, disconnected from social participation and fellowship, and inaccessible to social transmission of behaviors and attitudes" ("Being Autistic Together"). These perceptions and mischaracterizations make rhetorical identification between autistic and neurotypical people difficult because identification is often understood as resting on shared interests, connection, and fellowship. Additionally, successful identification between a rhetor and his or her audience often hinges on the construction of *ethos*—the way that a speaker, writer, or communicator characterizes himself or herself to an audience. Conveying goodwill, trust, and credibility to audiences, while inviting connection, is typically crucial to character formation and identification. For people with autism, traditional approaches to *ethos* are problematic, in part because of stereotypes about autism that may affect audience reception and inhibit identification, but also because of the pervasive neurotypical foundations of *ethos* in rhetorical theory.

The common stereotype that autistic people cannot connect with other people is a misconception that inhibits *ethos* construction and constrains possibilities for identification with audiences. As Dawn Prince-Hughes describes in her memoir of autism, "The public at large tends to hold in its collective consciousness a certain manifestation" of autistic deficits such as "impairments in the use of nonverbal, expressive gestures," often believing that "all people with autism are by definition incapable of communicating, that they do not experience emotions, and that they cannot care about other people or the world around them" (*Songs of the Gorilla Nation* 28, 31). She contends that "in many cases quite the opposite is true" and describes a "significant number of autistic people who care deeply about all manner of things" and who are "profoundly emotional" (31). Although Prince-Hughes, Sinclair, and many other writers with autism resist the assumption that autistics are cold, unemotional, or unconnected, stereotypes of the asocial person with autism persist. If people with autism are assumed to be emotionless, apathetic, and unconnected to the world around them, audiences are less likely to be receptive to their efforts at *ethos* formation. Taken to the extreme, the logic of the asocial stereotype of autistic people casts them as devoid of character, unable to craft personas in rhetorical exchanges, and incapable of developing a social connection or identifying with their audiences.

Simply put, autistic people are seen as *ethos*-less when viewed through a narrowly medical or pathological lens. Leo Kanner, who first described autism among eleven individuals he studied, identified his subjects as having an "*inability to relate themselves in the ordinary way to people and situations from the beginning of life*" and called these inabilities "*inborn autistic disturbances of affective contact*" (41, 50, emphasis in original). Hans Asperger, another early researcher, concluded, "The autist is only himself (cf. the Greek word *autos*) and is not an active member of a greater organism which he is influenced by and which he influences constantly" (38).[1] In these perspectives, people with autism are portrayed in ways that disqualify them from rhetorical agency in that they are described as unable to possess character, to respond to rhetorical situations, or to connect or identify with others. Recent definitions of autism and diagnostic criteria contain echoes of these early characterizations. *The Diagnostic and Statistical Manual of Mental Disorders IV*, for example, describes people with autism and Asperger syndrome in deficit-driven language through "failure to develop peer relationships," "lack of social or emotional receptivity," and "impairment in social interaction" and the use of nonverbal behaviors. Dominant current theories of autism, although contested, characterize people with autism as having "mindblindness" or the inability to imagine other people's states of mind (Baron-Cohen). Applied to rhetorical contexts, this emphasis on failure, lack, and deficit characterizes people with autism as unable to craft character or to connect with a wide range of potential audiences neurotypical and neurodiverse.[2]

Specifically, descriptions of autism focused on the failures of autistic people in the areas of social communication, interaction, and receptivity inhibit *ethos* construction according to traditional rhetorical techniques. The rhetorical tradition outlines an implicitly neurotypical approach to *ethos* formation, providing models for character development for people who think, act, and communicate in exclusively neurotypical ways, rather than the neurodiverse ways of people with a range of cognitive differences such as autism. It is a given in the rhetorical tradition and contemporary rhetorical theory that *ethos* resides in a rhetor who connects easily to an audience, is in tune with the minds of others, and who can reciprocate feelings easily.[3] For people with autism, these assumptions do not fit the complex experience of presenting oneself as a rhetor in a largely neurotypical world.

Ethos as it is usually understood in the rhetorical tradition demands a neurotypical rhetor whose outward persona aligns with cultural and social norms.

Ancient understandings of *ethos* emphasize "the conventional rather than the idiosyncratic, the public rather than the private," and "the virtues most valued by culture" (Halloran 60). Isocrates, for example, discussing selections of topics that contribute to *ethos* formation for the rhetor, recommends against causes that are "devoted to private quarrels" and instead advises one to show support to topics that are "great and honorable, devoted to the welfare of man and our common good; for if he fails to find causes of this character, he will accomplish nothing to the purpose" (*Antidosis* 276–79; trans. Norlin 339). Successful construction of character depends on cultivating a persona that is public, social, and receptive. Anyone who "wishes to persuade people will not be negligent as to the matter of character," and "words carry greater conviction when spoken by men of good repute than when spoken by men who live under a cloud." Being isolated, private, or unreceptive weakens one's *ethos* in this perspective.

Ethos as it is traditionally understood also excludes many autistic rhetors because it assumes a seamless correspondence, based on neurotypical expectations, between internal and external states of mind. For Isocrates, "the power to speak well is taken as the surest index of a sound understanding, and discourse which is true and lawful and just is the outward image of a good and faithful soul" (253–56; 327). In this formulation of *ethos,* outward projections of ability, such as speaking well, indicate internal good sense, good character, and a "sound" understanding. Internal and external thoughts align in this approach to *ethos* projection: "for the same arguments which we use in persuading others when we speak in public, we employ also when we deliberate in our own thoughts." It is assumed that one's internal thoughts mirror one's external character and no tension exists between one's individual and social selves.

This seamless alignment initiates a transactional process through which character is habituated through interactions with audiences. According to Isocrates, the orator "will select from all the actions of men . . . those examples which are the most illustrious and the most edifying; and, habituating himself to contemplate and appraise such examples, he will feel their influence not only in the preparation of a given discourse but in all the actions of his life" (276–79; 339). A process of normalization through habit is the result of this transactional relationship, as the rhetor is expected to come to resemble the most illustrious and edifying of a society's examples. Neurotypical expectations undoubtedly shape what kinds of actions are deemed "illustrious" and "edifying," which in turn influences and normalizes the development of the rhetor's character. As Michael Hyde explains, "genuine character development" in Isocratean terms is dependent on "one's 'natural' capacity to use language" and means that "the orator is necessarily both a student and a teacher of the dynamics of civic responsibility" (xv). Inhabiting these dual roles of student and teacher requires not only a rhetor who easily shifts between roles, authority, and responsibility but also one who reflects the norms, virtues, and values of his culture.

For autistic rhetors, ancient models of *ethos* formation, including current legacies and iterations, clash with real-life experience. In resistance to the expectation that successful *ethos* formation reflects a seamless internal and external state of mind, Uta Frith describes people with autism as struggling to "predict relationships between external states of affairs and internal states of mind" (*Autism* 77). As Sinclair describes, for autistic people, experiences of being with other people are not so seamless or "clear cut," consisting of a range of difficult situations, including "being intruded on by other people wanting us to engage with them, when we don't share that desire; being interested and curious about other people, but finding them confusing and overwhelming to be around; trying to engage with other people, and having frustrating and unsuccessful encounters; managing to engage 'successfully' with other people, and finding ourselves drained and possibly even damaged as a result of what we had to do to 'succeed'" ("Being Autistic Together"). The costs of successful *ethos* formation and identification for people with autism can be high; to varying extents the rhetorical tradition and its legacy are inaccessible for autistics seeking to form *ethos* and initiate identification.

A transactional process between rhetor and audience based on neurotypical assumptions also characterizes the Aristotelian approach to *ethos*. Aristotle advises that "it adds much to an orator's influence that his own character should look right and that he should be thought to entertain the right feelings towards his hearers; and also that his hearers themselves should be in just the right frame of mind" (*Rhetoric* 1377b25–29; trans. Roberts 59). Rhetor and audience come to resemble and even mirror each other in this transactional and normalizing relationship. Elements of *ethos* such as looking right, sounding right, and being in the right frame of mind are highly determined by cultural norms and subject to a process of normalization. Being able to exhibit Aristotle's elements of a strong *ethos*—"good sense, good moral character, and goodwill" (1378a9–10; 60)—almost always means demonstrating neurotypical ways of acting, expressing, and communicating. Aristotle's advice to exercise the "right management of the voice" including volume of sound, modulation of pitch, and rhythm contributes to giving audiences the "right impression of the speaker's character" and shapes an implicitly able and neurotypical speaker (1403b10–35; 119). Some people with autism may experience difficulty adjusting their pitch, volume, and rhythm of speech to neurotypical standards, a characteristic that may disqualify them from good delivery or strong *ethos* in an Aristotelian perspective.[4] Looking right and sounding right mean contributing to conforming to audience expectations based on normalcy and neurotypicality.

Legacies of Isocratean and Aristotelian approaches to *ethos* in contemporary rhetorical theory often uphold implicit neurotypical assumptions. As Sharon

Crowley and Debra Hawhee explain, in ancient Greece character formation also included "the community's assessment of a person's habitual practices," meaning that "a given individual's character had as much to do with the community's perception of her actions as it did with her actual behavior" (167). For contemporary rhetors, much the same is true: a community's perception of one's behavior often shapes a given individual's *ethos* as much as any action or behavior of the individual. For people with autism, who exist in a largely neurotypical world, autistic behaviors and actions are often misunderstood, devalued, and judged by neurotypical conventions, rendering *ethos* construction challenging. As Melanie Yergeau explains, part of understanding both autism and neurotypicality as rhetorics involves "*calling attention to normalized discourse patterns frequently portrayed as desirable and ideal*" as well as to neuroatypical discourse patterns that are frequently devalued ("Autism and Rhetoric" 489). James S. Baumlin, exploring Aristotle's influence on modern rhetorical criticism, explains that in Aristotelian *ethos,* "the rhetorical situation renders the speaker an element of the discourse itself," including elements such as "the rhetor's physical presence and appearance, his gestures, inflections, and accents of style, [which] are all involved in acts of signification" (xvi). As Heilker and Yergeau suggest, autistic gesture, inflection, style, and presence are typically not understood by neurotypical audiences ("Autism and Rhetoric"). This gap in understanding creates challenges for autistic rhetors attempting to cultivate *ethos* and for audiences who attempt to be receptive to an autistic rhetor. As with expectations for physical presence, audiences also presume a certain kind of cognitive presence. When dominant interpretations of autism characterize people with autism as "mindblind," or unable to imagine other people's mental states, neurotypical audiences are likely to assume that autistic rhetors cannot carry out the basic elements of *ethos* construction and rhetorical identification, including the ability to connect to others and to imagine how they may be feeling or thinking.

Traditional Isocratean and Aristotelian approaches to *ethos* and their legacies are inaccessible for people with autism seeking to craft character. For rhetors with autism, achieving a seamless and transactional relationship with audiences is often challenging because the inner shape of one's character may not easily match an outer persona. Jim, a contributor to a collection on the experiences of autistic students in the academy and beyond, describes the discrepancy between internal and external "projections" of his character in his definition of what autism means for him: "It means I don't receive and process information in the same manner as other people, not that I am stupid. It means I don't share the general neurotypical population's innate receptive and expressive communication skills; it doesn't mean I am unable to have feelings and emotions or am unable to share those emotions with others" ("Jim" 67). People with autism may not possess a neurotypical sense of "innate" receptivity and expressiveness such as that modeled in Isocratean and Aristotelian traditions and legacies of *ethos*. This does not mean that people with autism are not receptive or not expressive but that they may receive and express information differently. They may express and receive sensory perception or emotion differently, which in turn may affect the volume, pitch, and rhythm of the voice. People with autism may share emotion and establish their characters differently, making the idealized transactional relationship between rhetor and audience evolve in unexpected ways.

People with autism may also mediate between a sense of individual and social self in ways different from those supported by traditional and contemporary methods of *ethos* construction. As Baumlin, applying Richard Lanham's notions of the self, notes, tension between a social self and a central self characterizes competing approaches to *ethos* in ancient and contemporary theory (xvii). For example, *ethos* translated etymologically as "character" would "seem to describe a singular, stable 'central' self," and *ethos* translated as "custom" or "habit" would "describe a 'social self,' a set of verbal habits or behaviors, a playing out of customary roles" (xviii). People with autism, often communicating in a largely neurotypical world, frequently experience tension between singular and social notions of self. Prince-Hughes, for example, describes the energy she exerts to translate between neurotypical and autistic codes, noting, "I am glad that I am so successful at appearing normal (whatever that is), but I also wish at times people knew how hard I work at it" (*Songs of the Gorilla Nation* 2). In contrast to the alignment of internal thoughts and external representation of those thoughts to an audience that Isocrates and Aristotle imagine, Prince-Hughes notes barriers: "I am always aware of a moving sort of glass between me and the world, my present and my heritage, what is seen and what is not seen and only felt" (4). Temple Grandin, in an often-cited phrase, describes her experience in neurotypical culture as one in which she feels like an "anthropologist on Mars" (Sacks). Rather than the seamless combination of individual and social selves projected by both ancient and contemporary approaches to *ethos*, people with autism such as Prince-Hughes and Grandin struggle to reconcile their personas with a wider culture that is unfamiliar with or misinformed about the differences of autism.

In the face of this struggle, people with autism such as Prince-Hughes and others frequently depend on a rhetorical strategy of *ethos* construction based on touch to shape new methods of constructing character in resistance to a neurotypical rhetorical tradition. As is explored in this chapter, by using touch, rhetors with autism form new approaches to *ethos* construction, forging identifications with other people and resisting pervasive stereotypes about the experience of autism. People with autism demonstrate themselves as connected with their audiences and purveyors of goodwill and good character by forming habits based on touch and embodied in *mētis*, a practice traditionally understood as cunning intelligence.

After exploring the habits of *ethos* more in-depth, I redefine *mētis* as a tactile relationship of embodied cognition between people and their environments that supports a method of character formation not based on traditional notions of ability and neurotypicality. Through the construction of a *mētic ethos* based on the habits of touch, autistic rhetors forge identifications with a wide range of people. By exploring *ethos* through the tactile rhetorical practice of *mētis,* I revise *ethos* for neurodiverse rhetors and the audiences with whom they engage.

The Tactile Habits of *Ethos*

Understanding *ethos* construction as a tactile process of habituation performed by autistic rhetors forwards recent explorations of autism as a richly embodied and diverse experience. In many cross-disciplinary conversations, autism, portrayed previously as a primarily cognitive condition, is being understood through embodied, sensory, and affective registers. Melissa Park, exploring the power of folk neurology in occupational therapy environments, argues for valuing the "embodied pleasures," especially sensory and movement-based perceptions, that people with autism experience and share with others ("Beyond Calculus"). Anne Donnellan, David Hill, and Martha Leary, drawing on reports from autistic self-advocates, neurologists, and child development researchers, argue for exploring the importance of sensory and movement differences in autism, explaining that understanding somatic difference in autism is crucial to encouraging communication, relationship, and participation ("Rethinking Autism"). In his reflections on parenting an autistic son and his study of online autistic writing, James C. Wilson suggests that certain difficult-to-understand autistic behaviors such as self-injury may be the exercise of an "embodied language" (*Weather Reports* 73). Another parent of an autistic son, Ralph James Savarese, calls attention to "the closeness that autism has engendered" to describe family relationships in a roundtable discussion of caregivers, siblings, and other family members (Savarese and Savarese "The Superior Half"), a characterization that resonates with Chloe Silverman's argument for understanding autism through the use of love as an analytic tool. This closeness of autism—realized through sensory perception, movement, caregiving, pain, or pleasure—can also be understood through the sense of touch, particularly when touch is harnessed to form rhetorical *ethos.*

Despite frequently being stereotyped as uncaring, alone, and asocial, people with autism often find alternative ways to connect rhetorically with their audiences, depending on the sense of touch to craft *ethos.* Lucy Blackman, a woman with autism who spent much of her early life uncommunicative, offers an account of how the sense of touch figured into her first use of language via facilitated communication—a system involving a computer or letter board and a facilitator who provides physical support to the hands, arm, or wrist.[5] Blackman describes her experience with Rosemary Crossley, her facilitator:

> My partner placed her hand on mine again . . . I pressed single keys through the little holes, and suddenly my finger felt again as if it were a willing agent of my mind. In stabilizing my hand Rosie had given me another gift. My hand and my mind were connected. The eyes are in some ways a periscope between the mind and the outside world. As with a periscope, their effectiveness depends on compensating for wave movement and also the judder of on-ship activity. However my body, my mind and my senses had never interacted in quite the same way as my sisters' so I had some difficulty in knowing exactly what I could expect of all my bits and pieces, especially if I were trying to focus on, and also use, my hand in conjunction with abstract thought. The steady touch on my own hand and forearm somehow made me bring it into focus, and at the same time feeling the point of contact gave me an accurate measurement as to the distance between my fingertip and my sensation of that touch. (83)

Blackman draws attention to her distinct relationship to the sense of touch, as she locates herself in relation to her facilitator and the technology she uses. She differentiates her experience with touch from her neurotypical siblings' experiences and explores in detail how her hand, mind, and senses connect through her facilitator's touch.[6] Crossley's touch locates Blackman's senses in her own body and facilitates a habit that fosters movement toward the keyboard.

Touch, however, also poses problems for Blackman. Reflecting on her discomfort with physical stimulation, Blackman probes her complicated relationship to touch and communication: "If affection in the form of cuddles and kisses cause discomfort and pain . . . how on earth does one develop interaction which might compensate for not interacting to speech and glance?" (Biklen et al. 146). She answers herself, writing, "I worked out when I was writing my autobiography . . . that being touched in my skin and skeleton made my body aware of cause and effect. This created a clumsy crutch on which to hang expression. It does not mean that I changed, though I did make rational and analytical sense of what I was typing" (147). The sense that Blackman makes hinges on the active negotiation of touch in daily life and communication. Touch functions as a "crutch on which to hang expression" for Blackman, as she negotiates her body's boundaries and its relationship to cause and effect in the world around her. The result is progress in understanding how to express herself in writing.

Other people with autism also credit a complex negotiation with the sense of touch to progress in their daily lives and, eventually, toward advancements in communication. As Grandin writes, "Tactile stimulation for me and many autistic children is a no-win situation. Our bodies cry out for human contact but when contact is made, we withdraw in pain and confusion" (Grandin and Scariano *Emergence* 36). In *Thinking in Pictures* she reconsiders, however, writing: "Even though the sense of touch is often compromised by excessive sensitivity, it can sometimes provide the most reliable information about the environment for people with autism" (*Thinking in Pictures* 65). Prince-Hughes explores a complex negotiation of touch as well, writing, "I love the feel of having my scalp massaged and my arms tickled lightly more than traditional forms of physical contact," but she notes, "I startle and must fight rage when someone touches me unexpectedly" (*Songs of the Gorilla Nation* 2). Similarly the facilitated communication user Tito Mukhopadhyay explains that as a child, like Prince-Hughes and Grandin, he resisted physical interaction. He describes a lack of bodily awareness: "It took me many years to realize that I have a body. . . . Even this day sometimes I feel that I am walking without legs" (Biklen et al. 137). The education specialist Douglas Biklen, who interviewed Mukhopadhyay about facilitated communication, asked him to explain this particular relation to his body, remarking, "Given your sensitivity as a young child to being picked up, it is ironic that touch became so important to your learning" via facilitated communication (Bilken et. al. 138). Mukhopadhyay replied, "To think about it, I recall that I learnt every skill through the touch method. I have a problem with imitating any movement by looking at people performing or by mapping my body according to the instructions given to me. . . . Different skills need different time to practice depending on the feeling of awareness on that part of the body. Sometimes I feel my legs better than the hands. But I needed my mother's help to learn the tricycle. She had to manually push my legs because I could not do the movement. It needed some practice before I could ride it independently" (138). Mukhopadhyay, like Blackman and many other people with autism, possesses a complicated and even contradictory relationship to touch. Learning to negotiate touch—exploring the discomfort and advantages of it—is the key to Mukhopadhyay's learning. By practicing touch and habituating himself to the feel of his body in relation to others, he learns a method by which he can engage with the world. Grandin, Prince-Hughes, and Mukhopadhyay pursue active negotiation of touch, working with its benefits and its contradictions, to explore their relationships to themselves, other people, and the world around them.

People with autism such as Blackman, Mukhopadhyay, Grandin, and Prince-Hughes use touch rhetorically to form new embodied habits for situating themselves in their environments and in relation to other people, efforts which contribute to their active invention of *ethos*. As Crowley and Hawhee note, for the ancients, *ethos* was not "given by nature" but instead was "developed by habit" through the "practices in which [people] habitually engaged" (167). Aristotle, for example, conceived of a process of embodied habit in rhetorical training, advising that "[diligent attention] (*epimeleias*) and studies (*spoudas*) and exertions (*suntonias*) are painful; for these [too] are necessarily compulsions unless they become habitual; then habit makes them pleasurable (*ethos poiei hêdu*)" (Kennedy *On Rhetoric* 1370a4; Hawhee "Bodily Pedagogies" 149). A mixture of pleasure and pain accompanies this habit formation for Aristotle, as the nouns used—*epimeleias, spoudas,* and *suntonias*—connote a "force of intensity" (Hawhee "Bodily Pedagogies" 149). *Epimeleias,* in particular, with its root of *meletê,* is a "means of cultivating ê*thos,*" connoting "pain" in its plural form and "practice" and "exercise" when used to refer to rhetorical training (Walker 148; Hawhee "Bodily Pedagogies" 149). As Hawhee points out, this exercise of habit, typified by *meletê,* points to "a mutually constitutive struggle among bodies and surrounding forces" (150). The habits of *ethos* develop amid the context of multiple bodies and forces in tactile relation.

People with autism who negotiate the complexities of touch in their lives cultivate touch as a habit that drives the bodily production of *ethos.* Rhetors with autism such as the ones described in this chapter actively engage in the process of habit formation, using touch to situate themselves within their environments and to identify with others. Grandin, Prince-Hughes, Blackman, and Mukhopadhyay describe both pleasure and pain associated with the sense of touch as they learn to manage their own bodies and to communicate effectively with others. Touch operates for them as a way to form and reform certain habits in *ethos* construction, a process that requires intense attention, struggle, and concentration in relation to other bodies and forces. They practice a kind of training such as that described by Aristotle in *Nicomachean Ethics,* which possesses the potential to change or even become one's nature in *ethos* formation: "habit is easier to change than nature; for even habit is hard to change, precisely because it is a sort of nature" (qtd. in Hawhee "Bodily Pedagogies" 150). The complex habituation through touch that people with autism undertake enables character formations that resist the stereotypical asocial "natures" that autistic people purportedly possess.

This construction of character through the habits of touch is a mobilization of an *ethos* attuned to the material and spatial aspects of character. As Hyde points out, *ethos* possesses connections to materiality and space that are frequently neglected. *Ethos* in this context "refer[s] to the way discourse is used to transform space and time into 'dwelling places' (*ethos;* pl. *ethea*) where people can deliberate about and 'know together' (*con-scientia*) some matter of interest" (Hyde xiii). This "primordial" meaning of *ethos* predates its more familiar meanings of "moral character" and "ethics," opening the term up to "such dwelling places [that] define the grounds, the abodes or habitats, where a person's ethics and moral character take form and develop" (xiii). In the spirit of this material sense of *ethos,* touch can also be understood as a "dwelling place" for people with autism to form *ethos.* Autistic people form character in the "habitual gathering places" of touch—the spaces and places in which they dwell in their everyday lives, at work and in writing—often "knowing together" and developing selves in relation and amid multiple forces. In these dwelling places, people on the spectrum practice "being autistic together," as Sinclair puts it, as well as being autistic in a larger neurotypical culture, often seeking connection and identification. Although autistic rhetors often describe discomfort or uncertainty with touch, they force themselves to negotiate its complexities, an effort that, as I explore, enables many to form certain habits through touch that develop their *ethos* and strengthen opportunities for identification with others. I explore one such crucial habit formation through a rereading of the ancient rhetorical practice of *mētis,* which is a key rhetorical and tactile strategy among rhetors with autism.

The Tactile Habits of *Mētis*

The sophistic practice of *mētis* offers possibilities for understanding *ethos* as a habituated and tactile practice that draws on a wide range of diverse cognitions and forms of embodiment, regardless of traditional notions of ability and neurotypicality. In their book-length exploration of *mētis,* Marcel Detienne and Jean-Pierre Vernant, tracing *mētis* over ten centuries of Greek culture, define *mētis* as a "mental category" or a type of "wiley intelligence" evident in many different personas and operating in diverse contexts, ranging from "the forms of knowledge of Athena and Hephaestus" to technical objects such as "a hunting trap" or the "fishing net" (2–3).[7] Although Detienne and Vernant have difficulty defining *mētis,* they identify four core features, describing it as showing "the opposition between using one's strength and depending on *mētis*"; operating on "shifting terrain, in uncertain and ambiguous circumstances"; being "multiple and diverse"; and exhibiting itself as a "power of cunning and deceit" through tricks or traps (13–21). The first characteristic of *mētis,* its opposition to strength, demonstrates *mētis* as a particularly valuable embodied practice for people with disabilities, as Jay Dolmage has demonstrated in his exploration of Hephaestus.

Mētis is both a cognitive and embodied practice ripe for the uses of nonnormative rhetors. As a type of intelligence, *mētis* is "a way of knowing" that draws on a wide range of alternative cognitive strategies, including mental attitudes and intellectual behavior such as "flair, wisdom, forethought, subtlety of mind, deception, resourcefulness, vigilance, opportunism, various skills, and experience acquired over the years" (Detienne and Vernant 3). *Mētis* also includes the traits of a wide range of diverse bodies and embodied knowledge, including "the mastery of the navigator, the flair of a politician, the experienced eye of the doctor, the tricks of a crafty character such as Odysseus, the back-tracking of a fox and the polymorphism of an octopus, the solving of enigmas and riddles and the beguiling rhetorical illusionism of the sophists" (2). Recent retheorizations of *mētis* explored by Hawhee and Dolmage emphasize the embodied aspects of *mētis.*[8] Focusing particularly on the twisted limbs and curved movements of Hephaestus, the "crippled" god of *mētis,* Dolmage suggests that this example of "bodily difference and cunning thought can be reclaimed as a means of challenging physical and intellectual norms" (125). By focusing specifically on the inherent tactility of *mētis,* I build on and extend recent efforts by Dolmage, positioning *mētis* as a practice that revises the limits of *ethos* for rhetors with autism who use the sense of touch to communicate and construct *ethos.* In particular, understanding *mētis* as a bodily habit—a type of *hexis*—based on touch opens up *ethos* to the possibilities of habituated actions generated from relationships among bodies and their environments in contact.

Mētis, as both an embodied act and a form of intelligence, is primarily a manual practice based on touch. *Mētis* manifests in the manual "skills of a basket-maker, of a weaver, [and] of a carpenter" (Detienne and Vernant 2). Hand-wrought bonds through nets, weavings, traps, or other devices are the tactile productions of *mētis.* Manual actions drive *mētis* and produce its effects: "Bonds are the special weapons of *mētis.* To weave (*plékein*) and to twist (*stréphein*) are key words in the terminology connected with it" (41). A wide range of human, technical, and animal entities—often in physical contact—exhibit *mētis.* The body of the octopus, for example, with its "thousand interweaving arms," exhibits *mētis* because "every part of its body is a bond which can secure anything but which nothing can seize" (41). In the treatises of Oppian, "bonds, ropes, cords made from twisted willow and twisted snares" are the tricks of the trade, forming the "basic material" for hunters and fishers (41). The "good hunter" prepares for his hunt through the "art of binding" by making "willow withies," which are two to four strands of willow twisted together, joined end to end (41). This manual dexterity also characterizes certain personas displaying *mētis* in Greek culture. Hermes, for example, is known as "a living web of interweaving" because he is the "twisted or sly one," a "creature as mobile as the mime, *Strophios*" and his son, *Phlogios,* also a mime, who both "could imitate the most diverse living creatures with movements of their agile fingers and hands" (41). In Greek culture, these bodies and personas are inhabited with tactile properties, manual dexterity, and cunning flexibility.

The terminology and practices of *mētis* and the rhetoric of the sophists are deeply interconnected through touch. The sophists derived their name from the terms *sophos* and *sophia,* which in one meaning refer to skills in handicrafts (Kerferd 24.) The name of "Strophios," the mime who makes the shapes of countless animals with his hands, is related to the name "Strophaios" which is "the name given by the Greeks to the sophist who knows how to interweave (*sumplékein*) and twist together (*stréphein*) speeches (*lógoi*) and artifices (*mēchanaí*)" (41). Terminologically the tactile practices of *mētis* are linked to the rhetorical practices of the sophists. The fishing frog, for example, uses a ploy called a *sóphisma* to wriggle out and escape the lure of the fisherman's bait, matching the *mētis* of the fisherman. Plutarch calls cuttlefish, cousin of the octopus, *sóphisma,* because "they have a long, thin tentacle which they move very slowly to lure the fish" that they seek to catch in "living traps" (38). Plutarch and Aristophanes describe the *mētis* of animals as models for a similar *mētis* in hunters, fishers, and sophists. Even when animals such as the fishing frog, for example, seem to be caught, they triumph: "they have all the cunning of the sophist, the *poikílos* schemer 'who is never without a way . . . of escaping from difficulties'" (33). Among hunters, fishers, and sophists, the *mētis* of one's adversaries, human or animal, is an advantageous trait and a model for developing *mētis* for oneself.

These terminological associations signal deep connections between *mētis* and *ethos* formation. *Mētis* is a practice developed by habitual actions and reactions through a tactile system of repetition and response, a cycle that also includes character formation. As Hawhee describes in *Bodily Arts, mētis* is a *hexis,* a habit, in that it involves a bodily comportment of rhythm, repetition, and response that develops certain advantageous skills. In tactile encounters, these skills often develop regardless of ability or disability, in a transformative process between bodies. In hunting and fishing, for example, the *mētis* of the hunter and the fisher competes against and matches the *mētis* of the hunted and the fished. As Detienne and Vernant, drawing from Oppian, explain, it "is the *mētis* of the fish which obliges the fisherman to deploy an intelligence full of finesse" (33). A systematic cycle of outwitting and outsmarting directs *mētis,* as "hunters and fishermen must be capable of showing *mētis* superior" to their adversaries; they must have "more tricks up their sleeves than their victims" (33). Plutarch, in *On the Intelligence of Animals,* explains that octopus hunting develops a man's skill and practical intelligence. In fact the potential of *mētis* to affect the development of intelligence and *ethos* formation is so transformative that it leads Plato in the *Laws* to condemn the practices of "line fishing, the hunting of aquatic creatures, and the use of weels, the hunting of birds, and all forms of hunting with nets and traps" because "all these techniques foster the qualities of cunning and duplicity," which threaten his definitions of virtue and, by extension, good character (33). In these ways *mētis* operates as a type of *hexis*—a system of forming habit through tactile practice and response—that shapes the actions and *ethos* of people ranging from hunters and fishers to sophists and citizens. In contrast to the Isocratean or Aristotelian models of *ethos* formation based on normalizing transactions and mirroring between rhetor and audience, *mētis* models an *ethos* formation based on a transformative tension of give and take among a wide range of bodies and abilities.

The wide variety of tactile skills on which *mētis* draws invites alternative models of *ethos* formation drawn from diverse embodiments. The polymorphism of the octopus models a particularly *hexis*-driven *ethos.* Drawing on the flexible *mētis* of the octopus in his advice for *ethos* formation, Theognis counsels, "Present a different aspect of yourself (*epístrephe poikílon ēthos*) to each of our friends. . . . Follow the example of the octopus with its many coils (*polúplokos*) which assumes the appearance of the stone to which it is going to cling. Attach yourself to one on one day, and, another day, change colour. Cleverness (*sophíē*) is more valuable than inflexibility (*atropíē*)" (Theognis 213–18; trans. in Detienne and Vernant 39). The stiffness of "*atropíē* is strictly opposed to *polytropíē,* as immobility and rigidity to the constant movement of whoever can reveal a new face on every different occasion" (39). The ideal form of *ethos,* from the perspective of *mētis,* resides in someone "who can turn a different face to each person" (39). This form of *mētic ethos* encourages a different character for every situation. The supple arms of the octopus, who can take the shape of anything to which he clings, serve as a physical model for the cognitive dexterity needed for reshaping one's *ethos* for different audiences and situations.[9] A flexible, tactile reshaping of the octopus body models the diverse cognitions advantageous for the *ethos* of human rhetors.

Mētic Formations

People with autism who have complex relationships to touch in their lives often exhibit tactile *mētis* in their *ethos* and character formations. Prince-Hughes, for example, enacted an octopus-like *ethos* when she took a job at a zoo and began getting to know her other colleagues. After carefully observing the actions and communications of gorillas she worked with, she began mirroring their behavior in her human interpersonal encounters. Describing her colleagues, she writes,

> I believe they had started to care for me because I had begun to find ways to make it known to them that I cared for them and had good intentions. I cared about them as people, and after years of watching gorillas, I was learning how to tell them that. I knew to smile at them when I saw them. I knew how to put them at ease by sitting near them and to show interest in their lives by asking questions. I learned to show them a face that demonstrated sincerity and concern: I would consciously knit my eyebrows together and nod as they told me of their troubles. Like a gorilla, I would touch their shoulder to show them I was with them through their sadness and worry. When they told me something about their good fortune or relayed a funny story, I consciously laughed and brightened my face, squinting my eyes and turning my mouth up. (*Songs of the Gorilla Nation* 104)

Prince-Hughes displayed an octopus-like *mētis* in her conscious use of showing different faces to different people based on the situation. She put the habits she learned from gorillas into action. She used touch to connect with people and to communicate her concern. By creating faces and using touch, she visually and tactilely constructed an *ethos* based on goodwill, good character, and reciprocity among herself and her interlocutors. Prince-Hughes's *ethos* is flexible and rooted in a *mētic*-like strategy of responding to diverse situations and mediating between autistic and neurotypical modes of communication. In contrast to the neurotypical assumptions of *ethos* that undergird the rhetorical tradition, the practices of *mētis* that circulate in Greek culture through figures such as the octopus reveal alternative and neurodiverse approaches to *ethos* construction. In resistance to stereotypes that people with autism are cold, rigid, and overly literal, Prince-Hughes models an *ethos* that is warm, flexible, and interpretive.

The *mētis* of the octopus also offers redefinitions of *logos,* modeling habits for flexible ways of thinking that value a range of nonnormative cognitive styles and support *ethos* formations of neurodiverse rhetors. A sophistic approach to *logos* through *mētis* is cognitive and embodied. As Detienne and Vernant show, in Greek culture the *mētis* of the octopus is "a model for a form of intelligence: the *polúplokon nóēma,* intelligence 'with many coils'" (39). Furthermore, "this octopus-like intelligence is to be found" in two types of men, the sophist and the politician (39). They elaborate, explaining that "it is in his shifting speeches, his *poikíloi lógoi* that the sophist deploys his words of 'many coils,' *periplokaí:* strings of words which enmesh their enemies like the supple arms of the octopus" (39). In addition:

> the sophist is a master at bending and interweaving *lógoi*—at bending them since he knows a thousand ways of twisting and turning (*pásas strophàs stréphesthai*), how to devise a thousand tricks (*mēchanâsthai strophàs),* and, like the fox, how to turn an argument against the adversary who used it in the first place. Like Proteus, he can run through the whole gamut of living forms in order to elude the clutches of his enemy. The sophist is also a master at interweaving for he is constantly entangling two contrary theses. . . . Speeches interwoven like this are traps, *strephómena,* as are the puzzles set by the gods of *mētis,* which the Greeks call *grîphoi* which is also the name given to some types of fishing nets. With their twisting, flexing, interweaving and bending, both athletes and sophists—just like the fox and the octopus—can be seen as living bonds (Detienne and Vernant 41–42).

The *mētis* of the fox and that of the octopus, as models for the habits and intelligence of the sophists, emphasize the tactility of *logos.* Strings of words and interlocking theses are the products of bodies conditioned to be slippery, flexible, and nonnormative. In fables the fox is known as possessing words that "are more beguiling (*haimúloi logoí*) than those of the sophist" (Detienne and Vernant 35). These "caressing words" of the fox or the sophist are developed through tactile practices: "a *mētis* or a *dólos* . . . is woven, plaited or fitted together (*hupháinein, plékein, tektaínesthai*) just as a net is woven, a weel is plaited or a hunting trap is fitted together" (45–46). Through the process of this fitting together, the *mētis* of sophistic argument is "related to very ancient techniques that use pliability and torsion of plant fibres to make knots, ropes, meshes and nets to surprise, trap and bind" (45–46). For the Greeks, the "weaving [of] some scheme" in words resembles these ancient tactile techniques of binding, a process by which "many pieces can be fitted together to produce a well-articulated whole" (45–46).[10] These repeated actions of binding, weaving, and twisting are manual habits that form cognitively cunning rhetors.

A student with autism named Darius, a contributor to a book of personal essays by college students with autism, demonstrates the interweaving and slippery feel of interlocking *logoi* and theses in his writing, aimed at a neurotypical audience. In his essay's conclusion, he relates that he has polished his text to a certain extent for his audience but has refrained from too much editing that would have "ironed out many of [his] autistic thought processes" ("Darius" 41). He leaves "traces" of his cognition in his text, admitting, "I also feel a perverse pleasure in making you work for your information. There is some justice in the fact that you will have to adapt to my idiosyncrasies for a change, instead of me having to spend a lot of energy trying to figure out what you are about. Are you feeling tired and somewhat confused and do you have the disconcerting feeling that you have missed something, but do not know what and where? If you do, that may actually be the most valuable effect of this essay on you. It will make you appreciate the energy we have put into tuning in to you" (41–42). Darius exhibits an octopus-like *mētic* intelligence "of many coils" in his refusal to normalize his writing and thought processes. The feelings of tiredness, confusion, and depleted energy he evokes from his audience are a particularly productive kind of *mētic* trap in which neurotypical readers experience the challenges that many people with autism endure on a daily basis. Through this untraditional approach to the appeals of *logos* and *ethos,* Darius makes connections with a neurotypical audience by devising a way in which he can approximate some elements of the experience of autism for them, creating a partial but meaningful identification. While the "perverse pleasure" that he gains from this exercise is not typically understood as an act of goodwill that increases one's *ethos,* Darius arguably achieves a greater aim of *ethos* construction: he invites his neurotypical audience to feel with him the struggle of presenting oneself and of making connection. Darius takes a risk in his *ethos* construction but achieves potential connection through a display of his own *epimeleias*—diligent attention to the struggle, including the pleasure and pain, of communicating in a neurotypical world.

The tactile practices of *mētis* provide models for valuing how people with diverse cognitions, especially people with autism, shape *ethos* in nonnormative ways. In contrast to a rhetorical tradition based on neurotypical approaches to *ethos* that assume a seamless relationship between rhetor and audience and an innately receptive and expressive rhetor, *mētis* offers models of *ethos* formation for rhetors who may have to learn how to communicate across both neurotypical and neurodiverse audiences and who may initially struggle to connect to their audiences, whether because of cognitive differences or audiences' stereotypes of autism. In the remainder of this chapter, I turn to particular models of *mētis* and specific instances of neurodiverse rhetors forming *ethos* at the intersections between bodies. I explore first how Athena, the goddess of *mētis* who facilitates among bodies through her technical arts, offers a model for understanding the facilitative rhetorics of people with autism who often draw on a shared *ethos,* rooted in physical contact among bodies, to communicate via facilitated communication. I then explore how Hephaestus, the disabled god of *mētis* and technology, forms a model of *ethos* for people with cognitive differences who also depend on the physical intersections among bodies to invent *ethos.* In exhibiting the *mētis* of both these models, people with cognitive differences connect to other people physically and rhetorically, often forming identifications with audiences and interlocutors that resist the stereotypes of people with autism as asocial or unknowable. More broadly, these models of *mētis* enacted by the tactile rhetorics of Athena and Hephaestus also resist traditional notions of *ethos* as exclusively neurotypical.

Athena's Touch: Facilitative *Mētis*

The origins of *mētis* in the body of the goddess Metis herself and the myths of her daughter, Athena, illustrate an *ethos* that breaks free from the physical boundaries of the single body, forming character in touch-based relationships. Metis is named for Zeus's first wife, who possesses the potential to disturb traditional hierarchies, with her intelligence constituting "a threat to any established order" (Detienne and Vernant 108). Zeus attempts to contain Metis by "marrying, mastering and swallowing" her, possibly by tricking her into using her own ability of metamorphosis against herself (109). Metis, able to transform into the bodies of "a lion, a bull, a fly, a fish, a bird, a flame or flowing water," may have been persuaded by Zeus to turn herself into successively smaller animals until she was small enough to be swallowed (20, 109). Athena's birth coincides with these events as she takes on the characteristics of polymorphism, shared identity, and transformation.[11] At her birth Athena already contains within herself the bodies of many others.

In mythology Athena's *mētis* operates primarily as an intervening force, as she constructs a shared *ethos* between herself and the people she assists, often through a technology or technical object of her own design placed in the hands or bodies of those through whom she works. In the realm of warfare and on the battlefield, Athena uses her *mētis* in multisensory ways to change the course of battle. She lends Achilles her armor, which "makes a brilliant flame leap from his body," causing a distraction that alters the outcome of the battle: the "horses turn back and their drivers lose their heads" (Detienne and Vernant 181). Athena's battle cry also aids warriors. As Detienne and Vernant relate, "Like the nocturnal bird which follows her wherever she goes, the screech-owl (*glaúx*), which attracts and terrifies other birds with the fixity of its eye full of fire as much as by its cry, Athena triumphs over her enemies with the eye and also through the voice emitted by her weapons of bronze"; this "'voice of bronze' which Athena and her protégé sound out together . . . mercilessly transfixes [their] enemies" (182). With sight and sound, Athena speaks with and through the bodies of her protégés on the battlefield.

Athena also intervenes through her technological prowess in other realms, including agriculture, horse taming, and sea navigation, working through the bodies and particularly the hands of farmers, charioteers, and helmsmen. In each of these areas, she uses a technical and tactile *mētis* to shape and guide human affairs. Detienne and Vernant describe the value of these lesser-known abilities of Athena, particularly the way in which she "intervenes as a power endowed with *sollertia* [*mētis*], manual skill and practical intelligence" (178). She is known as the god whose "power of technology" manifests, in one instance, through her intervention in agriculture in the making of "the instrument, the technical object"—the farmer's plough—with which to harvest corn (178). Farmers, as the "handmen" of Athena, enact this practice of *mētis,* exhibiting "the skill and technical inventiveness which

Athena's facilitative mētis. *Courtesy of Staatliche Antikensammlungen und Glyptothek München, Munich, Germany. Photograph by Renate Kühling.*

can complement the activity peculiar to the goddess of cereals," Demeter (178).[12] Working through the hands of her intermediaries, Athena models inventive ways of combining human agency and technological potential.

Athena's *mētis* is a force of intervention that often brings human, technical, and even animal bodies together. Athena, known as *Chalinîtis,* "of the bit," demonstrates her *mētis* most distinctly through her work with horses. In an account given by the author of the *Periegis,* the "equine Athena" provides "assistance of the most practical kind" to Bellerophon in his victory over an unruly horse, Pegasus (Detienne and Vernant 187). In Pindar's detailed account, Athena demonstrates her "technical intelligence" wrought by "her very own hand" through the construction of a bit that tames the horse (187–88).[13] In this demonstration of assistance, Athena's mode of operation suggests a cooperation and facilitation of power among the three entities involved—human rider, horse, and bit. Athena's action is "doubly 'artificial'" not only because of her "technical adroitness" but also because of her strategies of facilitation: "her action intervenes from the outside, is of short duration, and is applied to a concrete object which does not belong to her" (198). Athena's power facilitates among multiple bodies, as "she always manifests herself 'at the side'" of those she is helping (198).

In the arena of ship navigation, Detienne and Vernant note how Greek thought "stress[es] the affinities" between ships and horses through the manual operation of the tiller and the reins (232). Similar to the way she works with Achilles, the farmers, and Bellerophon, Athena guides shipbuilders and navigators: "when Danaos is presented as the first man to build a ship he does so upon Athena's advice and with her assistance" (234). Again, Athena's technical intelligence is distributed; this guidance includes "the art of building as well as that of driving," with "wood-cutters, carpenters and ship builders," all craftsmen who "enjoy the protection and favour of Athena" (234–35). Consequently products such as "the ship and the chariot appear not only as manufactured objects but also as instruments of her activity" (237). Athena's guidance works even beyond the technical objects of these realms, as her influence works through multiple bodies to include all the people and objects involved.

Touch is the vehicle for Athena's facilitative *mētis,* as each of her interventions brings humans and technologies together in physical contact. Each of the arts of crafting armor, a plough, a bit, and a ship's navigation system depends on a tactile relationship between human agents and technology. This relationship is complicated and necessarily dependent on various factors that influence the holder of power in each case. Power circulates among the various human, technical, and even animal elements at different times. Athena's *mētis* mediates among the various interweavings of technical and human entities; her guidance and interventions recall "the Greek sense of *mēdesthai,* which is intimately connected with the intellectual activity of *mētis*" (Detienne and Vernant 230). This intellectual activity is tactile at its core and promotes a shared sense of cognitive ability, in which expertise with a technical object is mediated and distributed among at least two bodies. *Mētis* moves through diverse bodies and situations, facilitating, distributing, and circulating opportunities for action and agency.

Tactile Mediations of *Ethos*

People with disabilities who use facilitated communication and their personal attendants who provide physical support cultivate an Athena-like *mētis* in their shared rhetorical productions. The Greek sense of *mēdesthai,* especially Athena's mediating and facilitative *mētis,* is a model for rhetors who use touch and technology to communicate. Rhetors such as Blackman and others similarly enlist their facilitators, mediating, distributing, and circulating communication in ways that garner rhetorical agency. Blackman uses her facilitator's touch, which operates "at [her] side," to locate a sense of her body that she already inhabits but cannot fully control, tapping into her own sense of language that she already possesses but struggles to communicate to others. She describes Crossley's touch as a form of practical and technical assistance that provides a gateway for her: "the way this large, kind, noisy person touched my shrinking hand and anticipated in her own where my finger wanted to go, as if she could feel the muscles and tendons start to contract, was itself a code breaker" (82). This code breaking of touch functions much like Athena's *mētis.* Touch, for example, quells the "judder of on-ship activity" for Blackman, similarly to how Athena's *mētis* guides the shipbuilder and the navigator (83). Blackman finds that Crossley's support gives her "control of [her] own hand in a partnership of body and will that most people take for granted" (82). This touch becomes part of a process of habituation, as Blackman conditions herself to use Crossley's touch to relate to others and communicate. As Blackman describes, "under her touch I pointed with what felt to me like practised ease" (80). When writing her first few sentences using facilitated communication, Blackman realized the progress she had made in communication: "I, who could not speak my own thoughts, use pronouns or make comments except with concrete naming words, blurted out, letter by letter, an observation that I now see is as accurate as it is typical of autism: IRUNTIMEMYWAYNOT YOURS!" (84). Blackman and Crossley exhibit a shared sense of *ethos* through the sense of touch, but in these powerful assertions, Blackman retains control of her own *ethos* and self-presentation.

Other users of facilitated communication, breaking out of the mold of singular, neurotypical *ethos,* too exhibit an Athena-like sense of *ethos* facilitated by touch. As described, Athena and her "protégés" were known for "sound[ing] out together" a voice on the battlefield (Detienne and Vernant 182). Mukhopadhyay uses a method of speech-making that depends on facilitation via touch in a context of Athena-like *mētis*. He relates, "I was helped by mother with pushes on my back so that I could use my sudden breaths and bring out the sound produced by my vocal cords. That was the beginning of my speech. I can now enjoy hearing my own voice although very few people can follow my speech" (139). Mukhopadhyay's communication facilitator, his mother, guides the production of his voice, but Mukhopadhyay controls his own voice. He emphasizes this negotiation, saying, "I am stressing on how to do and not what to do because no one should have the impression that I did not know what to do" (138). He knows what to do to speak and what to say but requires physical assistance forming the muscle movements necessary for rhetorical action. Upon reflection Mukhopadhyay realizes that he has habituated himself to all kinds of learning through this tactile, facilitative method: "Touch is always a big help when an activity is new for me. Only through practice and through the gradual fading of the touch the activity can be done independently. I needed to be touched on my right shoulder, for doing any new skill, be it soaping, be it eating or be it learning how to take my bath, or be it learning how to write. So I consider that the touch method is the vital step to speed up my learning skill" (139). Like Blackman, Mukhopadhyay uses touch as a vehicle for rhetorical action and claims ownership of certain messages in writing and speaking. Similar to Blackman's experience, Mukhopadhyay's *ethos* is one that starts out shared but then transitions to an *ethos* more focused on self-presentation following conditioning via a *mētic* process of habituation.

Sue Rubin, a facilitated communication user, identifies the physical support of a facilitator as a key component not only to her initiation into language but also to her development of *ethos:*

> In 1991 when I was thirteen and in an eighth grade severely handicapped class, the school psychologist and speech therapist contacted my mom and suggested trying facilitated communication [FC] with me. My mom had heard about the ouija board affect [*sic*] and had read negative things about FC. She said she didn't want them to try it, but they were so persistent, she agreed and they came to my house. I was as amazed as they were when I could spell parts of words with someone pulling back on my wrist. . . . The teacher typed with me at school everyday and my mom typed with me at home. I slowly progressed from single words to phrases and sentences and paragraphs. During this time it became apparent that I had been learning from the mainstream classes and my older brother's homework, and storing the information in my head. My brain was so disorganized it took months for me to be able to organize information and retrieve it. It took even longer for me to be able to clearly express original thoughts, not just short answers to homework. ("Facilitated Communication: The Key")

Rubin had been previously uncommunicative, but facilitated communication and the habits it develops operate as a "code breaker" for her as with Blackman. Like Blackman and Mukhopadhyay, Rubin has fashioned her *ethos* gradually through the process and habituation of facilitation through touch. Months of practice and reconditioning supported her development of the cognitive strategies that led to identifications with others and to "original" thoughts. She describes the process more in-depth and, like Blackman and Mukhopadhyay, emphasizes the advancements she has made in self-presentation that resulted from facilitation:

> When a person begins FC training she sometimes has a working brain and just needs a way to express herself. This type of person learns FC quickly. If the person is like I was and absolutely awash in autism, it takes a bit longer. We start with support at the wrist if the person can isolate a finger. We immediately move support up the arm so the person doesn't become dependent on a lot of support. The facilitator never guides the user, but rather pulls the arm back to a neutral position above the target whether it is a keyboard or pictures. The facilitator might also intermittently take a hand away so the person feels he is able to do it with minimal support. The facilitator starts with simple yes and no questions, then moves to predicable answers, then answers with a few choices, and finally open ended answers. The goal is not only to be able to type without support, but to think and generate original thoughts. ("Facilitated Communication: The Key")

As Rubin describes, the facilitator does not guide the user to certain words but instead acts as a mediator of the communication process, eventually guiding a user toward expression of original thoughts. Although she had been diagnosed at a young age with severe intellectual disability, Rubin used facilitated communication to earn high school and college diplomas. She has expanded on the other non-academic gains she has made, offering an example that shows a fully developed sense of *ethos:* "Facilitated communication has enabled me to form friendships with my staff who are my peers. . . . I interview potential staff members and decide who I will have . . . the adult agency hire. I also decide who I want to fire if I don't like the way they relate to me. The people who I decide should become my friends do so because I have friendship to offer them in return" ("Facilitated Communication: The Key"). In this cultivation of goodwill and character, Rubin forges identification through offering reciprocal friendship. Rubin also values "emotional support" through facilitated communication, writing, "As much as physical support is essential to independence so is emotional support. . . . With greater communication comes taking responsibility for our words. Facilitators can help ease our frustrations by keeping us focused and encouraging accurate typing. As much as I try to keep myself engaged, it is always helpful having some encouraging words or simple touches to refocus my attention" ("Facilitated Communication Enables").

Rubin shares the responsibility for her words with her facilitator, forming identifications and connections with her collaborators.

Rubin, Blackman, and Mukhopadhyay use facilitated communication and touch as a *hexis* for constructing *ethos* and forming identifications that are receptive and reciprocal. Each has developed an *ethos* that is traditional in the sense that each cultivates goodwill with others, forms relationships based on reciprocity, and establishes himself or herself as possessing good character, but the way they achieve this *ethos*, through touch, is nontraditional. In fact their nontraditional approach to *ethos* revises the many ways in which the construction of traditional *ethos* is normalizing and neurotypical. Most cogently, communicators such as Blackman, Rubin, and Mukhopadhyay shatter the universal ideal of *ethos* residing in a singular, neurotypical body and mind. Instead they offer examples of ways to form a shared *ethos* that originates and develops in two bodies, at the site of touch. In the spirit of the "dwelling places" that *ethos* engenders, where people can "know together," as Hyde describes, neurodiverse rhetors inhabit a form of the closeness of autism that Savarese characterizes, working in physical relation to others. Rhetors such as Blackman, Rubin, and Mukhopadhyay cultivate authority through their physical and emotional connections with their facilitators and other people, resisting stereotypes regarding the characters of people with autism. Physical contact focuses their awareness and facilitates connections with others through communication, reciprocity, and identification.

Traditional Aristotelian and Isocratean models of *ethos* do not support the type of facilitated *ethos* formation that rhetors such as Blackman, Rubin, and Mukhopadhyay exhibit, but Athena's style of *mētis* provides a model for understanding and valuing the rhetorical production of autistic rhetors who use facilitated communication. Athena, mediating and working beside others, can be seen as a potential model for understanding the physical closeness of autism in the context of *ethos* formation. Athena, whether working "at the side" of a warrior, a farmer, a chariot rider, a shipbuilder, or a navigator, guides and mediates through manual skill and the sense of touch. Rhetors with autism practice an Athena-like *mētis* when they enlist facilitators to work at their sides to aid them in producing their own rhetorics.

Hephaestus's Hands: Tactile *Mētis*

In contrast to Athena's *mētis* of mediation, Hephaestus's *mētis* is more direct and palpable. Whereas Athena's *mētis* manifests itself often through an art, practice, or way of doing something, Hephaestus's *mētis* manifests as a way of being and as a bodily attribute. Integral to the formation of blood and bone in Empedocles's theories, Hephaestus is the "crippled" or "lame" god whose differently formed limbs embody *mētis* in Greek culture. His disability is directly linked to his abilities in the technological arts. Through his disabled body, Hephaestus illustrates the key property of *mētis* as the triumph of traditional weakness over strength. In the case of Hephaestus, this means that his "deformed" limbs are advantageous, enabling his craftsmanship. As the god of fire and technology, Hephaestus uses *mētis* to fashion everything from the tools of his trade to a rudimentary wheelchair. As Dolmage shows, Hephaestus's disability can be seen "as that which allowed him to 'dominate the shifting, fluid powers such as fire and wind' in his work in the forge" ("Breathe upon Us an Even Flame" 121). Hephaestus was "the famed inventor, the trickster, the trap-builder, and machine-creator of Greek myth" (119). According to Dolmage, Hephaestus enables an understanding of "*mētis* as a rhetoric" that recognizes "the body as rhetorical" and values "bodily differences as meaningful and meaning-making" (119).

Accounts of Hephaestus typically emphasize the nonnormative shape of his lower limbs. Detienne and Vernant focus on the feet of Hephaestus, especially their morphology as related to animal models of *mētis* such as the seal, crab, or fish. The

Hephaestus's curved appendages. Return of Hephaistos to Olympus, Dionysus, Maenad. Black-figured hydria from Caere, Ionic-Greek (525 B.C.E.). Kunsthistorisches Museum, Vienna, Austria. Photo credit: Erich Lessing/Art Resource, New York.

curved appendages of these beings exemplify a way of moving that is "pliable and twisted" and "oblique and ambiguous as opposed to what is straight, direct, rigid and unequivocal" (Detienne and Vernant 46). Dolmage also focuses on the feet of Hephaestus, especially the curved, doubling, and divergent movements they model, to forge positive associations between disability and rhetoric, writing that Hephaestus's "outward-facing feet and his lateral thinking were allied, and both became a metaphor for *mētis,* the ability to move from side-to-side like a crab, as opposed to the forward march of logic" (125). The lesser-explored hands of Hephaestus also point toward new engagements with movement and logic, both metaphorically and literally, and gesture toward locating contemporary expressions of *mētis* among people with disabilities who engage critically with technology and their environments. By extending Dolmage's argument for understanding *mētis* as disability rhetoric, it is possible to understand the hands of Hephaestus as a specific example of the distinctly tactile advantages of *mētis* for neurodiverse rhetors.

The hands of Hephaestus align him with a technology of touch in which disability is advantageous. In mythology Hephaestus is associated with crabs, animal models of *mētis;* he is said to be the father of these "metal-workers of the sea" (Detienne and Vernant 268). Noting that "*karkínos,* which is the Greek name for the crab, is also the word denoting the blacksmith's tongs," Detienne and Vernant explain that "for the Greeks the image of this crustacean creature of the sea could not be dissociated from that of the instrument which is an appendage to the hands of the blacksmith and enables him to handle the glowing metal" (269). Hephaestus's forelimbs are compared to the lower extremities of his crustacean analogue: the crab has "pincers . . . to grip tightly while the feet are responsible for movement over the ground" (270). According to Aristophanes, "These claws are not for the sake of locomotion but serve instead as hands, for catching and holding; and that is why they bend in an opposite direction to the feet which bend and twist toward the concave side while the claws bend toward the convex side" (quoted in Detienne and Vernant 270). The claws of the crab are compared favorably to the hands of Hephaestus in the adjective used frequently to characterize Hephaestus, *kullós,* which means crooked-fingered, and recalls the pincers of the crab (271). The doubly divergent shape and movements of the forelimbs and lower limbs of Hephaestus and the animal models with whom he is associated represent "a synthesis of all the different directions, forwards and backwards, and left and right" that are possible in bodily movement (270). For Hephaestus's trade, this multidirectionality and flexibility are crucial in cognitive registers as well: "In order to dominate shifting, fluid powers such as fire, winds and minerals which the blacksmith must cope with, the intelligence and *mētis* of Hephaestus must be even more mobile and polymorphic than these" (272–73). As Hephaestus is the god of fire and metalworking, his curved hands are crucial elements in his profession.

Hephaestus's blacksmith tongs. The Kleophon Painter (Greek, Attic), Skyphos, side A, detail of Hephaistos, ca. 420 B.C.E. Toledo Museum of Art, Toledo, Ohio. Purchased with funds from the Libbey Endowment, Gift of Edward Drummond Libbey, 1982.88. Photo credit: Tim Thayer, Oak Park, Michigan.

Hephaestus's craft provides an example of *mētis* that bridges embodied and cognitive registers. Hephaestus's hands enable him to hold and handle the tools and materials of his craft—metals and tongs—more effectively, a skill that shapes his *ethos* and intelligence as a craftsman. Specifically, through the sense of touch, particularly in the shape of his curved and tonglike hands, Hephaestus models a tactile approach to using disability, including cognitive difference, to technical advantage. As Hephaestus is a metalworker and god of technology, his *ethos* is hands-on and direct; he shapes his *ethos* as actively as he molds the materials with which he works. Although previous explorations of Hephaestus's *mētis* have focused on his bodily movements, equally valuable are Hephaestus's divergent ways of thinking. In contrast to a traditional rhetorical model of *ethos* that is universalistic, neurotypical, and normalizing, Hephaestus offers a model of *ethos* that is flexible, neurodiverse, and situation-specific. His metalwork calls for ingenuity and flexibility in comportment and cognition. In fact his body and mind function in a complementary fashion, creating nonnormative approaches to technology in which disability is an advantage. Although his *mētis* differs from that of Athena, he too uses his technical knowledge and intelligence to facilitate partnerships with others, often in direct, hands-on ways. As Dolmage relates, in one myth "he builds two voice-activated tripods, what we would call robots, to help him with his work, and he befriends Cyclops, teaching him to work with fire as well" ("Breathe upon Us an Even Flame" 127). As he is a clever trickster and trap maker, Hephaestus's engagement with technology and the materials of his work is inventive, collaborative, and creative.

Hands-on *Mētis* and *Ethos* Formation

People with autism working in technical and professional contexts enact a Hephaestus-like tactile *mētis* in their constructions of *ethos*, exhibiting a hands-on style of habituating themselves cognitively to the people, environments, and animals with which they work. Prince-Hughes and Grandin, both of whom experienced a conflicted relationship to the sense of touch as children, crafted a tactile sense of *ethos* as they matured in their personal and professional lives.[14] Undiagnosed with autism until the age of thirty-six, Prince-Hughes spent much of her life as a child and young adult attempting to figure out how to fit into a world from which she felt completely disconnected. She characterizes her confusion with the world as similar to the experience of "looking through an often opaque glass" and began to see more clearly only while visiting a local zoo and observing gorillas (*Songs of the Gorilla Nation* 4). She elaborates, writing, "I am always aware of a moving sort of glass between me and the world, my present and my heritage, what is seen and what is not seen and only felt." Prince-Hughes materializes her initial disconnection and engagement with the world through the metaphor and materiality of glass. Glass, cages, walls, and enclosures—the tricks and traps of *mētis*—are also the elements of her experience of autism.

Prince-Hughes actively cultivates a sense of tactile *mētis* to construct strategies for negotiating a world in which she feels set adrift. After observing the behavior and communication of the gorillas behind the glass as a spectator, she obtained a job at the zoo, started making important connections with her colleagues, and began making sense of her life and her world. Prince-Hughes describes, "I found a way to go home through the glass—the glass of my reality as an autistic person, certainly, but even more I found a way through the glass of a common zoo exhibit. The first time I knew that the glass was moving was a day, like the many days I sat with the gorillas, when we began to know one another" (*Songs of the Gorilla Nation* 4–5). This knowing that Prince-Hughes experiences—her way of going through the glass—is a tactile and cognitive experience: "I knew the glass was moving when a gorilla touched me. A gorilla touched me, and I connected to a living person as I had never done before" (5).

Prince-Hughes experienced this transformative touch in her capacity as a zoo aide, while charged with the task of feeding strawberries to Congo, a five-hundred-pound male silverback gorilla. A zookeeper showed Prince-Hughes how to lay out strawberries for Congo safely, keeping her hands back behind the bars, but Congo ate up the berries faster than Prince-Hughes could lay them down, and contact occurred:

> And then, in an instant, it happened. We put our fingers down at the same time. His gigantic finger, black and leathery, soft and warm, rested on my own digit. We stared at our fingers, and neither of us moved. Finally, I looked up into his soft brown eyes. They were dancing in surprise.
>
> We stayed like that for what seemed like a long time, our fingers joining five million years of evolution and reaching out to bridge the gap of generations traveled. He leaned forward slowly until he was six inches from my face. I could feel his breath. His steady eyes peered into my soul, and he did not blink. I leaned forward and rested my forehead on the bars. Our faces were almost touching. We stared at each other, our fingers still together.
>
> I relaxed into his touch and his nearness. *This is what it is,* I thought. *This is what it means to love and be loved. This is what it is to touch and look at another person and feel its meaning. This is what it is to not be alone in the vastness of the space we hurtle through among the coldness and the dying. This is what it is to live,* I thought. (*Songs of the Gorilla Nation* 6)

When Prince-Hughes's hand touched Congo's, their touch inspired Prince-Hughes to cultivate an *ethos* among people at the zoo. She studied the gorillas every day, interacting with them and imitating them. Observing the gorillas, she wrote, "I felt like I was watching people for the first time in my whole life, really watching them" (93). Spending hours in front of their window and writing meticulously detailed reports, she learned, by watching the gorillas, how to identify and communicate with people, beginning to see "basic patterns of humanlike behavior and discernible personalities, first on one side of the glass, then on the other" (111). For the first time she connected with the zoo staff, feeling that they were working together productively. She remarked that after obtaining a job at the zoo, "I was surprised to look around and realize that I had communicated with people toward a common goal and that they understood me" (101). Prince-Hughes's tactile encounter with Congo directly resulted in her tuning into her environment and the people around her, connecting with others more reciprocally and forging identifications. After connecting with Congo and then her colleagues, Prince-Hughes suddenly realized, "We are all in cages of one kind or another" and decided, "I would *be* among people, no matter what the pain. And though it *was* painful, though it meant fighting a losing battle for the gorillas, though it meant that my way of being a scientist would be rejected by most, I would turn my prison into a temple. I would reach out" (130). Reaching out is both literal and figurative for Prince-Hughes. In the spirit of

Hephaestus's *mētis,* a nonnormative touch via the hands catapulted Prince-Hughes toward ways of thinking and moving that resulted in the cultivation of a nontypical but successful *ethos.*

In a different but complementary context, Grandin also exhibits *mētis* in her approach to crafting *ethos* in her professional and personal life. Like Prince-Hughes, Grandin used a moving barrier of glass to explore how she learned to interact with the world, illustrating a *mētis* of flexible movement and cognition. She explains how she struggled in the "social area" of her daily life because, as a visual-symbolic thinker, she did not yet have "a concrete visual corollary for the abstraction known as 'getting along with people'" (Grandin and Scariano *Thinking in Pictures* 36). She relates,

> An image finally presented itself to me while I was washing the bay window. . . . The bay window consisted of three glass sliding doors enclosed by storm windows. To wash the inside of the bay window, I had to crawl through the sliding door. The door jammed while I was washing the inside panes, and I was imprisoned between the two windows. In order to get out without shattering the door, I had to ease it back very carefully. It struck me that relationships operate the same way. They also shatter easily and have to be approached carefully. I then made the further association about how the careful opening of doors was related to establishing relationships in the first place. While I was trapped between the windows, it was almost impossible to communicate through the glass. Being autistic is like being trapped like this (36).

Like Prince-Hughes, Grandin describes herself as trapped behind the glass of her autism but also sees this glass as a moving trap that can be manipulated and used to reach through to make connections with others.[15] Both see passages through the glass, understanding the glass and their autism as ways through the trap as well. For Grandin, like Prince-Hughes, this passage toward identification begins with the sense of touch, as she relates careful manipulation of the windows to navigating social relationships.

Grandin also exhibits a Hephaestus-like *mētis* in her fashioning of a piece of technology to facilitate her interactions with others. After spending a summer on her aunt's farm and observing the workings of a V-shaped pressurized cattle chute used to hold and calm animals during vaccinations, Grandin decided to build a similar machine for herself. Finding its gentle pressure relaxing, she describes using it as "the first time I ever really felt comfortable in my own skin" and articulates that "the feeling it gave me was one that I needed to cultivate toward other people" (*Thinking in Pictures* 63, 82). Grandin makes a habit of using her squeeze machine regularly, increasing her ability to form identifications with other people.

Characterizing this experience in terms of closeness, she writes, "When I was in the chute, I felt closer to people. . . . Although [it] was just a mechanical device, it broke through my barrier of tactile defensiveness, and I felt the love and concern of . . . people and was able to express my feelings about myself and others" (Grandin and Scariano *Emergence* 100). This advancement in self-expression led to progress in written expression as well. In her reflections on the chute, Grandin makes direct connections to writing, explaining, "The squeeze chute gives me the feeling of being held, cuddled, and gently cradled. . . . This is hard to write down. Writing it down is a form of accepting the feeling" (Grandin and Scariano *Emergence* 109). She describes her experimentation with the squeeze machine, writing, "It is like a language of pressure and I keep finding new variations with slightly different sensations. For me, this is the tactile equivalent of a complex emotion and this has helped me to understand the complexity of feelings" (Grandin *Thinking in Pictures* 90). Grandin attributes the squeeze machine to her "great strides in communicating with people" and even to her increased sensitivity toward the animals with which she works (108). Once she built the squeeze machine, she began thinking of how the feelings she experienced in the machine circulate through the animals she interacts with as an animal scientist and handling designer.

Grandin crafts a professional *ethos* with the insight that touch affords, becoming more receptive and expressive with the animals in her work. Describing the relationship between thinking and feeling, she again characterizes this association in terms of physical closeness, writing, "I always thought about cattle intellectually until I started touching them. . . . When I pressed my hand against the side of a steer, I could feel whether he was nervous, angry, or relaxed. . . . Sometimes touching the cattle relaxed them, but it always brought me closer to the reality of their being" (Grandin *Thinking in Pictures* 83). Connecting her knowledge of the sensory experiences of autism to the field of animal behavior, she knows that the "reactions of an autistic child and a scared, flighty horse are similar" in that "both will lash out and kick anything that touches them," but the "application of physical pressure has similar effects on people and animals" and can help both overcome "fear of being touched" (83–84).[16] As an equipment designer, Grandin translates the feelings she receives from animals through her hands into a rationale for her designs, creating humane restraint systems for the slaughter of sheep, hogs, and cattle: "As a result of my autism, I have heightened sensory perceptions that help me work out how an animal will feel moving through the system" (Grandin and Scariano *Thinking in Pictures* n.p.). Like that of Prince-Hughes, Grandin's autistic *mētis* recalls ancient models of *mētis,* such as Hephaestus's flexible thinking and curved appendages and movements. Grandin takes into account the characteristics of animals such as their heightened visual sense and sense of touch to design more effective and humane slaughtering systems. She uses a curved handling system, for example, for cattle because it is attentive to their natural circling behavior, the perspective of their eyes positioned on the sides of their heads, and their preference for continual contact with each other.[17]

Grandin's use of a curved, circular animal handling system is a clear example of *mētis* in action. As Detienne and Vernant explain, the terminology associated with *mētis* often involved the roots *gu* and *kamp,* which denote the qualities of the circle, the curve, and the bond. These shapes are associated with Hephaestus's *mētis* because they symbolize "feet [that] are twisted round or are capable of moving both forwards and backwards" as well as *mētis* in general because they denote "whatever is curved, pliable or articulated" (46). Grandin's ways of thinking and moving, especially present in her hands-on experiences with animals and her circular designs for animal handling, are similarly the result of knowledge that is "pliable and twisted" and "oblique and ambiguous" rather than "straight, direct, rigid and unequivocal" (46).

People with cognitive differences who use facilitated communication, such as Blackman, Rubin, and Mukhopadhyay, and rhetors who communicate more traditionally, such as Prince-Hughes and Grandin, all construct a sense of *ethos* shaped by touch and the flexible thinking it inspires. Although the appeal of *ethos* has traditionally excluded people with disabilities, especially those with cognitive differences, rhetors with autism show themselves to be actively forming constructions of good character, goodwill, and good sense, resisting stereotypes of autistic people as unreachable or asocial. Through negotiating actively with the sense of touch, rhetors with autism form identifications with their audiences and interlocutors in personal, professional, and everyday situations. They inhabit the closeness that autism occasions and use touch as a dwelling place through which to craft a neurodiverse *ethos.* These approaches to *ethos* necessarily contain emotional elements, which further resist stereotypes of people with autism as emotionless, unempathetic, or unfeeling. The habits that neurodiverse rhetors develop through touch increase opportunities for identification and contribute to establishing bonds among other people with autism and with neurotypicals.

5

Grasping *Pathos*

Physical Disability, *Kairos*, and Proximity

Physical disability, especially when it is visually conspicuous, often evokes an emotional response in audiences that may be at odds with the feelings or message that a disabled person seeks to convey. In her memoir of living with a highly visible degenerative neuromuscular disease, Harriet McBryde Johnson writes that after throwing away her back brace and letting her spine "reshape itself into a deep twisty S-curve," she was "entirely comfortable in [her] skin" (*Too Late to Die Young* 1–2). Other people, however, are not. She is weary of "the glassy smile" and the "concerned gaze" she receives from clients in her law practice when she greets them in her power chair (1). She is tired of the unsolicited comments from strangers on the street who offer her admiration "for being out" when "most people would give up" (2). She explains, "Because the world sets people with conspicuous disabilities apart as different, we become objects of fascination, curiosity, and analysis. We are read as avatars of misfortune and misery, stock figures in melodramas about courage and determination. The world wants our lives to fit into a few rigid narrative templates: how I conquered disability (and others can conquer their Bad Things!), how I adjusted to disability (and a positive attitude can move mountains!), how disability made me wise (you can only marvel and hope it never happens to you!)" (2).

As Johnson suggests, audiences, especially nondisabled ones, often experience limited emotions in response to her presence. These limited emotional responses that Johnson identifies in her own experience signal a larger issue surrounding the rhetorics of disability in the cultural imagination. As disability studies theorists such as Rosemarie Garland-Thomson and G. Thomas Couser have explored, dominant paradigms of pity, nostalgia, and other forms of sentimentality frequently accompany visual rhetorics or narratives about physical disability.[1] Visual rhetorics of people with visible disabilities often "act as powerful rhetorical figures that elicit responses or persuade viewers to think or act in certain ways" (Garland-Thomson "The Politics of Staring" 58). As Johnson's frustrating experience suggests, the power to move audiences is often circumscribed because of stereotypical "stock figures" and "rigid narrative templates," and only a limited set of emotional reactions or connections typically develops between disabled rhetors and their audiences.

In rhetorical terms, disabled rhetors are often limited by a predetermined set of emotions or appeals of *pathos* based on audience expectations about the negative experience of disability. Emotions focused on pity, sympathy, and inspiration frequently position people with disabilities as victims of a personal tragedy and passive objects uninvolved in emotional meaning-making. Rather than an interaction in which *pathos* develops as an emotional connection shared by audience and rhetor, these limited emotions prioritize the emotions of the audience of nondisabled people and often position the nondisabled audience in control of the encounter. As Howard Sklar and Wendy Chrisman have shown, even when emotions such as sympathy or inspiration may be useful rhetorically to rhetors with disabilities, these emotions are often received by audiences in limited ways, not unlike the ways Johnson outlines. As Garland-Thomson describes in her exploration of the visual rhetorics of disability, the objective is often to "appropriate the disabled body for the purposes of constructing, instructing, or assuring some aspect of a putatively nondisabled viewer" ("The Politics of Staring" 59). In disability and rhetorical studies, theorists such as Elizabeth J. Donaldson and Catherine Prendergast as well as John Duffy, among others, seek alternative approaches to *pathos* in rhetorics surrounding disability and seek new terms for emotional engagement among disabled rhetors and their audiences.

Similarly, Johnson wants to start from a new place with her audience and her readers. To write against the "formulaic narratives" that elicit predictable emotional responses about disability—pity, hope, sympathy, inspiration—Johnson claims a new beginning with her readers (3). She writes, "Instead of letting the world turn me into a disability object, I have insisted on being a subject in the grammatical sense: not the passive 'me' who is acted upon, but the active 'I' who does things" (3). Among these things are practicing law, traveling, participating in disability activism, and most important, telling stories. For Johnson, storytelling is "a survival tool, a means of getting people to do what I want," such as the stories she tells in exchange for "getting people to drive my van" (3). This transactional partnership also shapes her message: "My tales are true, or nearly true, as true as memory allows. They evolve in telling. They shift focus and emphasis, depending on what the listener—an active participant in the story's creation—wants. There are questions, digressions, reactions . . . I may recount the same sequence of events over and over again, but each listener makes a new story. That's because storytelling itself is an activity, not an object. Stories are the closest we come to shared experience" (4). For Johnson, shared experience shapes her life as much as story does. Describing her daily life, she relates, "My experience is crowded and overpopulated; I rarely do anything alone" (4). Telling stories in close proximity to her audiences gives Johnson opportunities to rewrite the narrative templates of disability and form alternative emotional connections with her audiences. In rhetorical terms,

Johnson—her active "I" as a storyteller—attempts to resist the dominant paradigm of *pathos* focused on pity and inspiration that drives many messages about disability.

Johnson's resistance focuses on the sensual experience of disability. Describing her frustration attempting to convey to her audiences the enjoyment she experiences in her life, she writes, "I used to try to explain that in fact I enjoy my life, that it's a great sensual pleasure to zoom by power chair on these delicious muggy streets, that I have no more reason to kill myself than most people. But it gets tedious. God didn't put me on this street to provide disability awareness training to everyone who happens by" (2). Part of the tedium that Johnson experiences is a product of the predetermined emotional response with which audiences receive her. Johnson's "sensual pleasure" of zooming in her power chair, an experience she tries to communicate to her audiences, resists formulaic narratives of disability and encourages new emotional connections with audiences, and yet audiences are unprepared for this message. They are unaccustomed to hearing about the sensual pleasures of people with disabilities, and Johnson struggles to get her message across. Emotions that counter the typical expressions of pity, inspiration, and hope often fail to register with audiences.

In this chapter I explore the sensual pleasure and complexity of touch as a vehicle for new emotional connections and inventive forms of *pathos* possible between disabled people and their audiences. Specifically, I explore how the rhetorical practice of *kairos*—a principle of timing and space—intervenes in limiting forms of *pathos* in rhetorical encounters involving rhetors with physical disabilities and their audiences. I redefine *kairos* through special attention to the sense of touch, showing how *kairos* operates tactilely to create new emotional and physical connections among bodies in close proximity and contact. Instead of formulaic emotions such as pity and inspiration, which operate on a subject-object binary, the sense of touch engenders a partnership and reciprocal interaction of emotion among people with disabilities and their audiences, facilitating identification.

Emotion and the Proximities of *Pathos*

Pathos is intimately connected to the embodied experience of disability in ancient and contemporary understandings, ranging beyond connotations focused simply on pity and sorrow and toward more active and transformative potentials. The etymological origin of the Greek word *pathos*, associated with *paskhō, paskhein,* or *paschein,* refers to "some kind of suffering" as well as to the more capacious experience of "feeling," "passion," and "emotion" (Freeland 233; Liddell and Scott).[2] *Paschein,* as described in Aristotle's definition of perception, when translated literally, possesses an emotional component of *pathos* and "lends itself to an ambiguity of feeling, as an active undergoing rather than the passive experience of pain"

(Paterson 19). As Laura Micciche points out, emotion is a transformative and social concept for Aristotle, who defines the emotions as "those things through which, by undergoing change, people come to differ in their judgments and which are accompanied by pain and pleasure" (Micciche 11; *On Rhetoric* 1378a8, 121). Emotion, in this context, is a "tool for *doing*" and is "experienced *in relation*, between people within a particular context" (Micciche 11–12). Multiple bodies are involved in this transformation, which seems primed for the needs and abilities of disabled rhetors who use touch as a rhetorical strategy.

Although Aristotle's sense of *pathos* in perception is transformative, his exploration of *pathos* in rhetoric, from a disability studies perspective, is more limited, tending toward normalization on the basis of ability and disability. Expressing ambivalence about *pathos,* Aristotle implies that the most able orators and audiences should not even require emotional appeals. He warns, "It is not right to pervert the judge by moving him to anger or envy or pity—one might well warp a carpenter's rule before using it" (*Rhetoric* 1354a24–26; trans. Roberts 3). The best or most able audiences—ones who are rational and determine truth easily—are ideal because they are not swayed by their emotions.[3]

Aristotle also creates binaries in the audiences he characterizes, prescribing normalizing emotions for them based on stereotypical understandings of ability and disability. For example, considering the "various types of human character in relation to the emotions and moral qualities," he contrasts emotions such as "anger, pity, fear and the like, with their opposites" such as calmness, friendship, and esteem (*Rhetoric* 1388b31–32; trans. Roberts 84, 1378a22; 60). Ability and disability are correlated with these emotions and moral characters. Aristotle contrasts youthful and elderly types of characters, being attentive to differences in abilities and emotions that direct each: old men are "past their prime," whereas young men are "changeable and fickle . . . their impulses are keen but not deep-rooted, and are like sick people's attacks of hunger and thirst" (1389b13; 85, 1389a6–9; 84).[4] Aristotle also associates sickness with uncontrollable emotion: "people who are afflicted by sickness . . . are prone to anger and easily roused: especially against those who slight their present distress" (1379a15–18; 62). He continues, "Thus a sick man is angered by disregard of his illness," a tendency that illustrates also how he is "predisposed, by the emotion now controlling him, to his own particular anger" (1379a18–23; 62). In other words, people with sick or nonnormative bodies are controlled by their emotions rather than being in control of them. From a disability studies perspective, Aristotle's aligning of healthy bodies with controllable or positive emotions and sick bodies with uncontrollable or negative emotions demonstrates a hierarchical approach to ability and disability. As the disability studies theorist Tobin Siebers has explored, the assumption that people with disabilities cannot control their emotions still exists; disabled people are often prone to charges of narcissism ("Tender Organs").

Aristotle again uses bodily vulnerability—relative strength and weakness—as examples for inducing emotion and identification among people, often toward normalizing ends. He counsels, "When it is advisable that the audience should be frightened, the orator must make them feel that they really are in danger of something, pointing out that it has happened to others who were stronger than they are, and is happening, or has happened, to people like themselves" (*Rhetoric* 1383a7–11; trans. Roberts 71). Aristotle's advice, if applied to the context of disability, potentially falls under the "hope it never happens to you" category of formulaic templates that writers such as Harriet McBryde Johnson identify.

Alternatives to the normalizing relationships between ability and emotion that Aristotle describes are possible when bodies come together in close proximity. Bodies in contact—in physical and proximal relation to each other in time and space—possess the potential to transfer emotions that can foster connection and partial identification. One of Aristotle's most important principles of *pathos* is that a rhetor must identify "what frame of mind" an audience is in to move the audience toward an emotion (*Rhetoric* 1377b24; trans. Roberts 59). As Craig R. Smith and Michael J. Hyde have shown, in Aristotle's treatment of *pathos,* "the intensity of emotions can be described in terms of the nearness or remoteness of the objects of emotions inclusive of the personal relationships that stimulate them" (450). They elaborate, "The closer what one fears is in time and space, the more intensely one experiences that fear; the more remote the object of fear, the less intense is the experience of fear. . . . Thus, a speaker can 'move' the listener to more or less intensely felt states of mind by bringing the objects of emotions closer or removing them from the listener's temporal/spatial field of perception" (450–51). Emotional states are affected not only by objects in relation but also by bodies in proximity. The proximity of people, including their degree of relationship, further figures into the experience of emotions such as fear and pity.[5] Generally, "a slight received from one who is close, such as a relative, is more painful, and therefore more intensely felt, than a slight received from one who is remote, such as a stranger" (451). Being "close" in familial relationship is only one type of proximity of bodies. Aristotle also recommends a "safe distance" between individuals, especially among those with significant differences in power (*Rhetoric* 1382b23; trans. Roberts 70). Time, which puts distance between individuals involved in a situation, likewise affects emotions such as anger: "when time has passed . . . anger is no longer fresh, for time puts an end to anger" (1380b6–7; 65). For Aristotle, the physical relationship of bodies in time, space, and proximity contributes significantly to the effects of emotions.

From a disability studies perspective, a focus on proximity explains the limited range of emotions that often accompany messages about disability focused on pity and fear. Listeners are not "moved" very far from their preconceived notions about disability if they encounter only appeals based on fear or pity. When emotions are limited by stereotypical portrayals of disability, as Harriet McBryde Johnson suggests, interaction and emotional connection between rhetors and audiences are also limited. In effect, a rhetor with a disability, especially a physical disability, experiences resistance in changing the "frame of mind" of an audience—often pity, fear, or even anger—and moving them to new emotions regarding disability. If disability is understood only as a misfortune to be avoided, then the people with whom Johnson interacts on a daily basis are likely to hold back the wide range of emotional connections possible. Emotions such as pity and fear become techniques that distance people emotionally and physically. Johnson, as described, has tired of the glassy smile and concerned gaze from others, indications of people's objectification of her, rather than their willingness to interact substantially with her. She prefers storytelling in face-to-face encounters developed within a partnership or friendship, favoring the experience of bodies involved in proximal contact and the timing and space of conversations in a transactional context.

Aristotle's concept of *pathos* via proximity, although it draws on normalizing examples, offers transformative potential for rhetors with disabilities seeking to develop new emotional connections with audiences. People with disabilities, such as Johnson and others, use proximity to create alternative configurations of emotion, untied to traditional messages about disability. In oral storytelling Johnson excels, drawing from the emotions of her listeners who are in close physical proximity in time and space to her: "Easter lets me know when I've found a character's voice by raising her right hand with an emphatic 'Thank you!' With Dave, it's silence that tells me when the pacing is right. . . . When Mike says 'Wait a minute, let me get this straight,' he leads me to an angle I hadn't noticed" (*Too Late to Die Young* 4). As emotions circulate among bodies in close proximity, the timing of Johnson's stories is directed and even changed by her audience's response; this emotional element and partnership are things she misses in writing. In Johnson's storytelling, emotion becomes a "tool for *doing*," a transformation forged through physical relationships among people.

The Proximity of *Pathos* as Catalyst

Rhetors with disabilities use emotion to transform attitudes and messages about disability, often drawing on the energy of bodies in proximity. Like Harriet McBryde Johnson, Nancy Mairs, a woman with multiple sclerosis (MS), initially notes the difficulties of connecting emotionally to audiences about her physical disability. She begins her memoir, *Waist-High in the World,* by writing, "I cannot begin to write this book. I've made some stabs at it, pried out of my rubbly brain a few pages, always 'preliminary,' from time to time. But mostly I write letters . . . or read material at least tangentially related to my subject, or merely play solitaire on my computer until I try my own patience to the point of despair" (3). Feeling isolated and impatient, Mairs struggled to begin writing. A phone call from a woman named Jennifer, however, who had just read one of Mairs's essays on MS and recognized the symptoms from her own experience, jogged her out of this state, and Mairs focused her writing toward the possibility of rhetorical identification.

Mairs realized that Jennifer, worried she might have MS, did not need a diagnosis but instead needed simply a "safe audience" to listen to her (5). She understood this, she wrote, because "like Jennifer, I often need no more than someone to whom I can speak frankly about MS without being dismissed as a whiner (a distancing tactic often practiced by those in whom disability triggers unbearable anxiety)" (7). Talking with and listening to Jennifer, Mairs arrived at a course of action: "What I'm supposed to do about Jennifer, of course, is to write a book: one in which she can recognize and accept and even celebrate her circumstances, but also one that reveals to those who care about her what needs and feelings those circumstances may engender in her" (6). In talking with Jennifer, Mairs experienced partial identification and then recognized her audience. She defined a purpose in writing based on validating feelings and creating connection. The timing of Jennifer's call was crucial, arriving at a critical point in the beginning of Mairs's writing process and creating an affective connection that inspired communication. Although their bodies may not have been in close proximity, their voices were closer in time and space via the telephone, and Mairs's desire to write and to connect with others intensified. Proximity, as Mark Paterson explains, connotes "both the physical nearness of tactile contact as well as the metaphorical nearness of empathy" (153), and thus propelled Mairs to rhetorical action. The voice on the line enabled Mairs to envision expanding networks of audiences, first based on proximity but then reaching across distances—not only Jennifer but also those who care for Jennifer.

In fact, in resistance to the "distancing tactics" of many audiences, Mairs desires to create a closer emotional connection between herself and her audience. This connection means bringing multiple bodies together in contact in new ways. Mairs's use of *pathos* connects bodies both physically and emotionally, mobilizing a circulation similar to what Ellen Quandahl recognizes as operating through the role of multiple, interconnected bodies in Aristotle's rhetoric, in which "emotion is not a simple phenomenon of the individual body, but a complex phenomenon of attention, body, belief, and judgment that can both contribute to argument and deliberation and be influenced by them" (17). Mairs draws on bodies in contact and connection in this transformative sense, writing, "In a society that prates about, but seldom practices, communication, the craving to be listened to, heard, understood—which originates with the first terrified wail, the circling arms, the breast, the consolatory murmur—is hard to assuage" (7). In Aristotelian terms, Mairs and other rhetors with disabilities face challenges getting audiences "in the right frame of mind" for making appeals based on *pathos* and use the proximities of bodies as a strategy for moving audiences differently. Audiences may resist appeals they see as falling into the category of whining and may "hear" only messages that fit into the formulaic scripts of pity and inspiration, but they may "feel" a different message through the proximity of bodies in time and space. Mairs invokes a primal scene of communication in her description of the birth experience, with its sounds and touches, in order to make a new model of communication based on senses and sharing. Later she again invokes the role of touch at the beginning and end of life to ask new questions about embodiment and communication: "I doubt that any body, whether in trouble or out, can fully conceive a self without an other to stroke it—with fingertips and lips, with words and laughter—into being and well-being. Research has demonstrated that infants deprived of touch fail to thrive, and that blood pressure is lowered and spirits are raised in elderly people given pets to caress. If physical stimulation is wholesome—even lifesaving—at the extremes of life, why should we suppose the middle to be any different? Our bodies conceptualize not only themselves but also each other, murmuring: Yes, you are there; yes, you are you; yes, you can love and be loved" (49–50).

The sense of touch functions crucially in the new model of communication Mairs desires, as she seeks to develop her relationship with audiences on a new emotional level, beyond formulas of pity and whining. Bodies in close proximity circulate new emotions. Physical proximity such as the kind that Mairs and Harriet Johnson describe mobilizes emotion as a "tool for *doing*" and takes as a "starting point, the Latin root of emotion, *motere,* which means 'to move,' suggesting that a tendency to act is implicit in every emotion" (Jacobs and Micciche "Introduction" 3). This tendency to act, by "linking emotion with movement," necessarily "underscores the rhetorical nature of emotion as a mode of articulation by which thought and action are moved, or are always in flux, and are a source for moving others" (3). Multiple bodies in proximity and contact are crucial for this emotional movement. For Mairs, touch and its benefits catalyze rhetorical action, intervening in an especially challenging episode of writer's block caused by isolation and disconnection.

Other rhetors with disabilities also use touch as a vehicle for bringing more bodies into proximity and creating new connections with their audiences, especially after frustrating experiences with writing. Kenny Fries, who was born with nonnormatively shaped legs, uses touch to create new models for communication in his poetry. After describing many unfinished and discarded drafts of poems, Fries began to work through the fear he has of writing. He writes:

> Tonight, when I take off my shoes:
> three toes on each twisted foot.
> I touch the rough skin. The holes
> where the pins were. The scars.
> If I touch them long enough will I find
> those who never touched me? Or those who did? *Freak, midget, three-toed bastard.* Words I've always heard. ("Excavation" *Anesthesia* 146)

Touching evokes conflicted memories for Fries, as he remembers being both touched and untouched. Like Mairs, Fries uses a birth scene to create a new emotional connection between himself and his audience, turning to the sense of touch to start anew. Plumbing the depths of his experience with hurtful labels, he continues:

> *Disabled, crippled, deformed.* Words
> I was given. But tonight I go back
> farther, want more, tear deeper into
> my skin. Peeling it back I reveal
> the bones at birth I wasn't given—
> the place where no one speaks a word. ("Excavation" *Anesthesia* 146)

In contrast to the hurtful words he has "always heard," Fries delves into the skin and discovers in silence the potential for new possibilities.

Eli Clare, like Mairs and Fries, turned to touch to approach the difficult task of writing about disability after experiencing challenges in previous efforts. Clare relates, "For years I have wanted to write this story, have tried poems, diatribes, and theories" (19). Clare seeks to shed the "second skin" that has resulted from negative assumptions about disability: "In the gawking, the pats on my head and the tears cried on my shoulder; in the moments where I become someone's supercrip or tragedy: all those lies became my second skin" (130). In an attempt to "reach beneath the skin," Clare focuses on touch once he begins writing about his experience with queerness and cerebral palsy (137). After asking questions about where and how to start writing his memoir, he asserts, "We cannot ignore the body itself: the sensory, mostly non-verbal experience of our hearts and lungs, muscles and tendons, telling us and the world who we are" (129). Seeking to uncover his own language of the body, he theorizes an understanding of the "body as home, but only if it is understood that language too lives under the skin," as a means to address the loss generated from stereotyping, shame and regret (11). To reconstruct the body as home, he includes writing about his life at different ages—thirteen, eighteen, and ending with his entry into the queer community as an adult. He writes, "I think of the butch woman, once my lover, now a good friend. One night as we lay in bed, she told me, 'I like when your hands tremble over my body. It feels good, like extra touching.' Her words pushed against the lies" (134). Touch catalyzes writing and connects oneself to others, forming a way of pushing against and resisting assumptions, stereotypes, and loss.

Touch figures significantly in the writing process of rhetors with disabilities such as Fries, Clare, and Mairs, often with touch functioning to produce new models of emotional engagement with audiences. Touch brings diverse bodies into proximity, changing the time and space of how emotions are shared among rhetor and audience. Fries, Clare, and Mairs all experience difficulty beginning the writing process but find that the sense of touch facilitates their arguments and reflections about disability, which in turn fuels their emotional connections and identifications with interlocutors and audiences. They, along with Harriet McBryde Johnson, exemplify Aristotle's theory of *pathos* as influenced by the proximity of bodies, adding the sense of touch to bring bodies closer together and to facilitate emotional connection. The proximity of touch—a closeness in space and time—often catalyzes writing and contributes to forming new emotional appeals beyond the formulaic emotions of pity and fear.

Rhetors with physical disabilities who use touch as a rhetorical strategy to create alternative emotional connections with their interlocutors and audiences position themselves as valuable contributors to efforts aimed at revising understandings of *pathos* and affect. As Jenny Edbauer Rice explains, critical affect studies (CAS) include emergent inquiries such as "a more complex understanding of *pathos* (beyond emotion), increased attention to the physiological character of rhetoric, and a rethinking of ideological critique" (211). Relatedly studies of "neurorhetorics" and the "neuroscientific turn" in the humanities and social sciences position affect, physiology, emotion, and cognition in interdependent connection (Jack and Appelbaum; Littlefield and Johnson). Disabled rhetors' contributions to these inquiries reveal alternative constructions of emotion in the context of disability. Harriet McBryde Johnson, Mairs, Fries, Clare, and others use touch in writing, speaking, and communicating in face-to-face encounters to force audiences to rethink hierarchies of ability and disability and to form new emotional connections with audiences. Using touch, a sense that is physiological and psychological as well as affective and cognitive, they call attention to the productive experiences that disability offers, particularly the experience of the body and emotion in time and space.

The Contacts of *Kairos*

Kairos is a rhetorical practice that, like the appeal of *pathos,* brings diverse bodies together in time and space. *Kairos* intensifies and drives the proximities of *pathos,* modeling a rhetorical practice for bringing diverse bodies even closer together in contact through time and space. Ancient associations of *kairos* often reinforce hierarchical divisions between ability and disability, but revised theorizations of *kairos* can be understood as situating the practice as primed for the challenges and advantages of disabled rhetors, especially those seeking new emotional connections to audiences through touch.

Kairos is usually translated in rhetorical theory as "opportune or appropriate time," but from its first uses in Homer and Hesiod, the concept of *kairos* is nearly synonymous with "disability," indicating places of bodily vulnerability and impairment that are penetrable tactilely. In Homer's *Iliad, kairos* is described in relation to a vulnerable bodily space, as it "denotes a *vital* or *lethal* place in the body, one that is particularly susceptible to injury and therefore necessitates special protection" (Sipiora 2). Hawhee notes that in the *Iliad* these vulnerable places, in the adjectival form of *kairos, kairios,* include "where the collarbone parts the neck and the chest" and are associated with "the crown of the head where the first hairs of horses grow on the skull" (quoted in Hawhee *Bodily Arts* 66). The hairs on the head are extremely sensitive to touch; touch receptors at the base of hair follicles are particularly responsive to pain and pressure. The adjectival form of *kairos* also describes the place that ancient archers sought to hit by aiming at "an opening or series of openings" on bodily targets such as the vulnerable gaps in the body's protective skeleton (Onians 345; Hawhee 66). In Hesiod's *Works and Days,* another early use of *kairos* accompanies advice against overloading the axle of a horse-drawn cart, referring to the bodily vulnerability of a working animal that is injured or broken by too much weight (Paley 694). In these early definitions of *kairos,* vulnerable bodies collide in time, space, and weight, often reinforcing negative associations between vulnerability, disability, and weakness. A unique tactile element operates in these early uses in which *kairos* is associated with penetration, weight, and porousness.

In other ancient depictions, *kairos* is frequently a tactile practice, a physical intervention in time and space with a normative purpose. Ancient senses of *kairos* emphasize medical intervention, often through normalizing diagnostics or in the search for a cure. Durations of illness, growth, and normal and abnormal human development are described in *kairotic* terms in some Hippocratic treatises. It is advised, for example, that "*every* disease can be cured, if you hit upon the right moment [*kairos*] to apply your remedies" (quoted in Atwill 57). In ancient Greek culture and medicine, doctors were known for possessing a sharp sense of *kairos* that enabled them to intervene physically against disease. Both Pindar and the anonymous writer of *On Ancient Medicine* emphasize the *kairotic* elements of medicine. Marcel Detienne and Jean-Pierre Vernant explain that the Greeks compared disease to the shifting and changing sea, likening the knowledge of the helmsman navigating a ship to the abilities of a doctor. A doctor must be agile enough to respond *kairotically* to the many symptoms and changing nature of illness; "medicine is an art of the fleeting moment (*oligókairos*), and in which the opportunities for intervention are always critical" (Detienne and Vernant 312). The doctor, for example, is "*epikairótatos*," like the pilot of a ship on raging seas, who exercises his "grip of time," or knowledge of when to intervene physically based on experience (312). Being able to foresee the course of a disease allows him to "watch for the precise moment" of intervention and "seize the opportunity of grasping Kairos by the hair" (312). This grasping of the god Kairos's forelock of hair by the physician represents the ability to move quickly and appropriately, to catch the opportunity of a passing instant to cure. Touch in these associations embodies the ability to cure and normalize a body.

This strategic grasping is particularly embodied through the statue of the god Kairos, who forms the model of ability and good health. Crafted in the third century B.C.E. by the sculptor Lysippos, the athletic Kairos adorns the entrance to the gymnasium in which contests of strength and ability were fought. Kairos is the exemplar of able-bodiedness, exhibiting a muscular body primed for fast movements. Depicted with wings on his heels, he is quick on his feet and able to respond quickly to changing situations. In a hymn by the poet Posidippos underneath the statue, Kairos is described as "stand[ing] on tip-toe," "ever running," and "fly[ing] with the wind," emphasizing his able-bodiedness through fast movement (*Anth. Pal.* 16.275; *The Greek Anthology* 325). His sense of touch enables him to grasp rhetorical opportunity. He is sharp and quick-witted, characteristics associated with tactility, in visual and verbal representations: his strong arm reaches across the center of the marble relief; he grasps a razor in one hand and balances a set of scales "as a sign to men that I am sharper than any sharp edge" (325). In addition he possesses a forelock of hair on his face "for him who meets me to take me by the forelock," grasping opportunity (325). The body of Kairos also represents the missing of opportunity passing by once it is gone, since with his winged feet and lack of body hair he explains that none "will take hold of me from behind" (325). This sense of movement is rooted in able-bodiedness, tactility, and agility. A quick, sound body represents a quick, responsive mind able to grasp rhetorical opportunities. The body of Kairos exhibits this ability through his skillful negotiation of touch and encourages others to do so as well in the grasping of opportunity.

In current rhetorical theory, *kairos* retains its ancient associations and is usually understood as a rhetorical concept depending on a sense of movement, often tactile, based on able-bodiedness. Both Eric Charles White and John Poulakos draw on the ancient abilities associated with *kairos* to redefine the concept for contemporary rhetors. White, calling *kairos* a "radical principle of occasionality" in which the "living present" spurs rhetorical invention (161), draws on the tactile elements of *kairos*: "In archery, [*kairos*] refers to an opening or 'opportunity' or,

Kairos's able body, holding a razor. Relief with personification of Kairos. Museo di Antichità, Torino, Italy. Courtesy of Archivio Soprintendenza per Beni Archeologici del Piemonte e del Museo Antichità Egizie.

more precisely, a long tunnel-like aperture through which the archer's arrow has to pass. Successful passage of *kairos* requires, therefore, that the archer's arrow be fired not only accurately but with enough power for it to penetrate. The second meaning of *kairos* traces to the art of weaving. There it is the 'critical time' when the weaver must draw the yarn through a gap that momentarily opens in the warp of the cloth being woven. Putting the two meanings together, one might understand *kairos* to refer to a passing instant when an opening appears which must be driven through with force if success is to be achieved" (13).

White's *kairotic* rhetor draws on the able weaver and powerful archer's skills, including the dexterity, force, and strength necessary physically to perform the grasping of the *kairotic* moment through which the yarn is woven or an arrow is fired. Poulakos's *kairotic* rhetor also acts ably as "both a hunter and maker of unique opportunities, always ready to address improvisationally and confer meaning on new and emerging situations" (J. Poulakos *Sophistical Rhetoric in Ancient Greece* 61). The "always ready" *kairotic* rhetor resembles the body of the god Kairos: responsive, agile, quick on his feet, and able.

Despite this emphasis on ability, recent revisions of *kairos* such as those by White and Poulakos, among others, can be seen as supporting the efforts of disabled rhetors because they acknowledge rhetorical situations in which numerous forces and bodies come together in proximity and physical contact. Debra Hawhee, for example, focuses on athletes who possess *kairos,* but she also opens up *kairos* to a wider range of bodies through Gorgias's concept of somatic *logos,* a formulation indebted to his teacher Empedocles's theories of pores and effluences. For both Empedocles and Gorgias, all "bodies and souls, like bronze and silver, were porous entities that allowed effluences and other substances (words, fire) to pass through" each other (Hawhee *Bodily Arts* 79). In Gorgias's rhetoric, "speech itself becomes a mobile body, shot through with *kairos,*" a result of "a mingling of porous, effective bodies" and operating as a "performative demonstration of the way *kairos* fold[s] so neatly into bodies in motion" (83). All bodies, traditionally able or not, become *kairotic* through connecting, circulating, and folding into one another.

Ability is not simply a given in retheorized definitions of *kairos,* and bodies are not necessarily independent entities but come together in new ways. Scott Consigny's definition of *kairos,* for example, calls attention to ability and skill, noting a sense of contingency: "*kairos* is an opportunity the rhetor discerns and helps to bring about during the course of the agon, given his perspective and abilities or skills" (*Gorgias, Sophist and Artist* 87). Similarly, Margaret Price's definition of *kairotic* space, which pays attention to the "less formal, often unnoticed areas" in which "knowledge is produced and power is exchanged," recognizes the importance of "in-person contact" in these scenarios (*Mad at School* 60–61). Bodily difference, including disability and ability, affects *kairos. Kairotic* bodies—interconnected, interdependent, and contingent—are not uniformly able.

The movements and circulations among bodies in contact that White, Poulakos, Hawhee, Consigny, and Price underscore exemplify the tactile roots of *kairos.* In addition to the tactile practices of *kairos* exhibited by Kairos the god, doctors, and helmsmen, there exist strong associations between *kairos* and specific practices of tactile arts such as weaving.[6] Richard Broxton Onians, considering the spatial aspects of *kairos* as well as its notions of time, illuminates the contact among the bodies of *kairos* in a wide semantic field. *Kaîros,* as a key term of weaving, signifies the place where threads are attached to the loom and the process of fastening the threads to the loom. In an array of other *kair-* related terms, *kairōma* indicates a web, *kairōstis* and *kairōstris* indicate the woman who weaves, and something that is woven is *kairoseōn.* As Onians explains, "*Kaîros* was evidently in some sense the warp or something in the warp, something to do with the 'parting' of its threads. It is generally taken to mean the row of thrums which draw the odd warp-threads away from the even, making in the warp a triangular opening, a series of triangles, together forming a passage for the woof " (346).[7] *Kairos* is a locus of bodies and materials in contact and movement and often a space in which "warps" or nonnormative locations are productive.

The movements of opening and closing that characterize *kairos*—the opening of a gap in weaving or archery, the closing of that space when a thread or arrow passes through—bring bodies and materials together in physical contact. Barrett Shaw, in *The Ragged Edge,* draws implicitly on a sense of *kairos* similar to the movements that Onians describes to convey the "feel" of the experience of disability, writing with emotional and tactile imagery: "It is hard to unravel the tangled, knotted ball of the disability experience—isolation and differentness versus a common identity. . . . This book attempts to weave a rough but strong cloth from these gnarled strands, to give the feel of the disability experience. Such a cloth would not have a neat, finished selvedge, but a ragged edge" ("Introduction" xii). Similar to the odd warp threads that Onians describes in the semantic field of *kairos,* Shaw's sense of the disability experience seeks to bring bodies together in nonnormative ways in space, movement, texture, and time.

This tactile movement of *kairos* is Empedoclean in form and function, modeling diverse and nonnormative approaches to space and timing. The basic principle of Empedocles's worldview, in which four root elements pass through channels of bodies, objects, and materials, is based on the assumption of movement in time and space via tactile and physical intermingling and mixing. This tactile sense of *kairos* as movement resists the normative sense of *kairos* in ancient medicine and the able body of the god Kairos. Tactile *kairos,* especially in weaving, occurs when time and space are somewhat warped, making a space and time for different formations of materials, bodies, and forces. In fact this warping is advantageous, forming a "passage" based on unevenness and difference, resisting senses of *kairos* in Hesiod and Homer that situate *kairos* purely in negative, traditionally disabling contexts. *Kairos* in this alternative sense is a rhetorical action precipitated on tactility, an action in time and space that happens among various, diverse, and nonnormative bodies in movement and physical contact.

Disability *Kairos*

A tactile sense of *kairos* is valuable for disabled rhetors because it provides models for productive action in the highly changeable, dynamic experience of disability. Disability theorists often employ an implicit sense of *kairotic* timing and space in order to explain the radical temporality, instability, and changeability of bodies that disability can occasion. Michael Bérubé writes that "any of us who identify as 'nondisabled' must know that our self-designation is inevitably temporary, and that a car crash, a virus, a degenerative genetic disease, or a precedent-setting legal decision could change our status in ways over which we have no control whatsoever" (viii). Lennard Davis notes that "all it takes is the swerve of a car, the impact of a football tackle, or the tick of a clock" to propel one into a different bodily status (*Bending Over Backwards* 4). Disability, whether acquired or inherited, is an experience in which a "normal" or expected sense of the body in time and space may change dramatically. People with disabilities use tactile *kairos* to display inventive rhetorical responses to the experience of disability. Disabled people often produce rhetorical action in creative and spontaneous ways, illustrating examples of *kairotic* action carried out by multiple bodies in physical contact in time and space.

As Carolyn Miller explains, *kairos* has several meanings in rhetorical theory and practice. People with disabilities frequently resist the first of two meanings of *kairos* in which *kairos* is associated with "propriety or decorum" and a "principle of adaptation and accommodation to convention, expectation, predictability" (C. Miller "Foreword" xii). A normative sense of order guides this notion of *kairos* because "violation of [this] order, failure to know the *kairos* and observe its propriety, will result in rhetorical, aesthetic, and even moral failure" (xii–xiii). In simply deciding to speak or write, a disabled rhetor may break an existing and tacit sense of order, decorum, and propriety that stipulates that people with disabilities are not "appropriate" rhetors because of perceived physical, cognitive, or emotional differences. People with disabilities are particularly constrained by the *kairos* of order, convention, and decorum because disability has so often been viewed as a disruption of order, a violation of convention, and an insult to decorum, an attitude exemplified in codes such as the "ugly laws," which sought to erase visible disability from public view in the late nineteenth and early twentieth centuries (Schweik *The Ugly Laws*). When Mairs, Fries, and Clare feel reluctance in writing their stories, they experience the effects of this static sense of *kairos*. The sense of accommodation and adaptation in this valence of *kairos*, in the context of an ableist world, places the burden of accommodation on the person with a disability to conform to nondisabled expectations or to keep disability hidden and out of view.

Rather than conforming to decorum or propriety—in effect adapting to what has come before—a different meaning of *kairos* looks ahead and generates invention. People with disabilities such as Harriet McBryde Johnson, Mairs, Clare, and others mobilize a use of *kairos* that is "uniquely timely," "spontaneous," and "radically particular," finding ways to be "creative in responding to the unforeseen, to the lack of order in human life" (C. Miller "Introduction" xiii). In speaking or writing, disabled rhetors improvise and adapt often in radical particularity, generating meaning in unprecedented and emerging contexts. As Miller explains, this dynamic, alternative use of *kairos* is "timely action" in which *kairos* "will be understood as adaptive, as appropriate, *only in retrospect;* it cannot be discovered within the decorum of past actions" (xiii). White highlights the transformative and unique valences of this more dynamic sense of *kairos,* noting that Gorgias's *kairos* "stands for precisely the irrational novelty of the moment that escapes formalization" (20). For Consigny, Gorgias's *kairos* means being able to "'leap' beyond the limits of *logos* into the unpredictable, chaotic flux of an irrational Becoming" (*Gorgias, Sophist and Artist* 45). Nonrational aspects of *kairos* also inform Dale Sullivan's reading of a "*kairos* of inspiration" as "poetic frenzy" or "divine madness," ("Kairos and the Rhetoric of Belief" 319) which Hawhee, drawing from the Empedoclean theory of pores and effluences, considers "somatically as the act of breathing in, or a commingling of momentary elements" (*Bodily Arts* 71). From a disability studies perspective, any of these senses of *kairos* may be present when any disabled rhetor undertakes a rhetorical action. These approaches to *kairos* typify the experience of disability, valuing nonnormative and nonrational rhetorical actions made in unique circumstances by bodies in contact.

The experience of disability inspires *kairotic* action because it often necessitates a nonnormative approach to the passage of time. *Kairos* is, in Bernard Miller's terms, "the way we measure our lives on the basis of critical events that disrupt the normal sequence of *chronos,*" or quantitative time (169). As Harriet McBryde Johnson, Mairs, and others show, the experience of disability affords the opportunity to act *kairotically* and to respond to unique and particular situations, conferring new meaning and responses. Mairs relates the temporal instability of disability in positive terms, resisting a normal or quantitative (*chronos*) way in which time should pass and exercising a sense of disability *kairos* in response. Because of MS, she describes how she has "had to learn to take satisfaction in stasis," and she reports, "I can entertain myself for long moments" by observing others (*Waist-High in the World* 37). Describing a sense of time aligned with *kairos,* she writes, "The languid, pensive state in which I now live much of the time has calmed me and expanded my contentment immeasurably. Because my slightest gesture requires effort now, I must focus on each moment, without much regard for past mistakes or the future's threats or blandishments" (37). Mairs aligns herself with the concept of crip time—a perspective in disability culture in which time does not adhere to normative measures (Gill; Zola). She is content with her newfound ability to live in the moment, a *kairotic* strategy occasioned by her disability.

Disability, whether genetic or acquired, is a "critical event" and is often accompanied by unprecedented circumstance that disrupts the normative passage or sequence of time in lived experience, often in diverse and productive ways. Simi Linton, cataloging the relatively recent "coming out" of disabled people into public and civic life, describes how time, space, and movement are diverse in the experience of disability: "We are everywhere these days, wheeling and loping down the street, tapping our canes, sucking on our breathing tubes, following our guide dogs, puffing and sipping on the mouth sticks that propel our motorized chairs. We may drool, hear voices, speak in staccato syllables, wear catheters to collect our urine, or live with a compromised immune system" (*Claiming Disability* 4). This collective portrayal of people with disabilities occupying space and time differently in public complements Linton's descriptions of personal and professional examples of disabled people moving in proximities that encourage emotionally productive experiences, such as the intimacy of learning how to dance using a wheelchair for the first time with a dear friend or the enjoyment of participating in the annual Society for Disability Studies dance with a group (*My Body Politic*). Using *kairos,* people with disabilities meet "the challenge . . . to invent, within a set of unfolding and unprecedented circumstances, an action (rhetorical or otherwise) that will be understood as uniquely meaningful within those circumstances" (C. Miller "Introduction" xiii). With *kairos,* particularly its tactile qualities, rhetors develop productive nonnormative and nonrational responses to the disability experience. Touch intensifies the radical potential of *kairos* for disabled rhetors, putting diverse bodies and emotions in contact.

Kairotic Touch

Often rhetors with disabilities mobilize a sense of improvisational *kairos* to connect with their audiences emotionally, using the sense of touch as a vehicle for this expression. Harriet McBryde Johnson, who excels at storytelling when her audiences are in close proximity, also uses touch *kairotically* in more explicitly rhetorical situations such as debates. When she participated in a debate with the Princeton philosopher Peter Singer, Johnson used *kairotic* touch to inspire new emotional connections among her audiences, even as she struggled to connect or identify, however partially, with Singer on certain levels.

Singer is well known for his stance on infanticide for severely disabled infants and his arguments for the liberation of animals. Johnson describes his philosophy, attempting to explain: "He insists he doesn't want to kill me. He simply thinks it would have been better, all things considered, to have given my parents the option of killing the baby I once was, and to let other parents kill similar babies as they come along, and thereby avoid the suffering that comes with lives like mine and satisfy the reasonable preferences of parents for a different kind of child" (*Too Late to Die Young* 201). Based on the principles of utilitarianism, Singer argues that infants who lack supposedly essential characteristics of personhood—rationality, autonomy, and self-consciousness—should be killed to avoid suffering and to give parents opportunities to have other children who will flourish. Some of Singer's criteria for personhood are based on time: awareness of "one's own existence in time" and the "capacity to harbor preferences as to the future" (202). Johnson has trouble wrapping her head around Singer's "tight string of syllogisms," and her initial response to him elicits more questions than anything else (201). Most broadly, she wonders, "What does it take to be a person?" (202). A particularly difficult question for Johnson is "How can he put so much value on animal life and so little value on human life?" (203).

Johnson first met Singer when he was invited to an event in her hometown to lecture. Johnson was dispatched by the organization Not Dead Yet, which spearheads opposition to legalized assisted suicide of people with disabilities. When she initially encountered Singer at the College of Charleston, she almost did not want to engage him in any type of contact or discourse, a hesitation that was an implicit attempt to forestall the possibility of any rhetorical identification. She conveys this reluctance via a description of her hesitation regarding whether or not to offer Singer her hand to shake. When Singer extended his hand to Johnson, she paused, thinking, "I shouldn't shake hands with the Evil One," but then, remembering rules of southern politeness, she reconsidered: "I give Singer the three fingers on my right hand that still work," and identified herself as a representative for Not Dead Yet (204). Extending her hand, she formed an embodied rhetorical response based on the sense of touch, hoping to unsettle Singer. After the handshake, Johnson "want[s] to think he flinches just a little" (204), but she realized that what "stands out in memory about that first meeting is Singer's apparent immunity to my looks, his lack of the visible discombobulation I expect, his immediate ability to deal with me as a person with a particular point of view" (204). The prospect of rhetorical identification, however fraught, incomplete, and uncertain, was ushered in with this handshake. This tactile rhetorical encounter foreshadowed the more powerful rhetorical strategies via touch that Johnson would marshal at their later encounter at a debate in Princeton.

Even though Johnson agrees with Not Dead Yet's stance—that Singer's views are "so far beyond the pale that we should not legitimate them with a forum"—Johnson represented Not Dead Yet in a question-and-answer session in Charleston and then accepted Singer's invitation to a public debate at Princeton (*Too Late to Die Young* 208). Although other disability activists may choose a purposeful silence to express disagreement with Singer or even to refuse him a forum, Johnson realized that for her, silence is not the only answer to Singer's views. She knew, for example, that by accepting Singer's invitation to Princeton, she had the chance to connect with different and more diverse audiences and perhaps forge identification with these audiences: "It seemed like an unusual opportunity to experiment with modes of discourse that might work with very tough audiences and bridge the divide between our perception and theirs. I didn't expect to straighten out Singer's head, but maybe I could reach a student or two" (Johnson "Unspeakable Conversations"). Johnson felt called to rhetorical action by the force of speech itself, especially in the chance to create new relationships with audiences. Listening to Singer in Charleston, for example, she wrote, "Even as I am horrified by what he says, and by the fact that I have been sucked into a civil discussion of whether I ought to exist, I can't help being dazzled by his verbal facility. He is so respectful, so free of condescension, so focused on the argument, that by the time the show is over, I'm not exactly angry with him" (*Too Late to Die Young* 206). Instead she was more angry with the audience: "I am shaking, furious, enraged—but it's with the big room, two hundred of my fellow Charlestonians who have listened with polite interest, when in decency they should have run him out of town on a rail" (206). It was important for Johnson to accept Singer's subsequent invitation to debate in Princeton because she felt strongly about changing audiences' perceptions of Singer's ideas and creating new emotional connections among her own audiences.

When she arrived at Princeton, however, she found little productive audience involvement and few opportunities for emotional involvement and identification with her audience. She lectured and received questions from the attending students that were "fairly predictable" (215). She felt that in the question-and-answer session following the debate on infanticide, many of the discussion topics and responses were ones she could have "emailed in," until a student attempted to engage Johnson in a difficult topic. She called on a young man in the room, who asked her, "Do you eat meat?" Johnson replied, "Yes, I do," and the student countered, "Then how do you justify—," but Johnson interrupted, explaining, "I haven't made any study of animal rights, so anything I could say on the subject wouldn't be worth everyone's time" (215). Johnson cut off this student's question and any exploration of the topic by pleading ignorance and by noting the limits of time. When she and Singer discussed the question later, Singer reframed the question from his student: "What he wanted to know is how you can have such high respect for human life and so little respect for animal life" (217). Johnson exercised a rhetorical reversal, replying, "People have lately been asking me the converse, how you can have so much respect for animal life and so little respect for human life" (217). But when Singer started to answer, she abruptly severed the dialogue again, saying, "Look. I have lived in blissful ignorance all these years, and I'm not prepared to give that up today" (217).

When preparing for the public speech at Princeton, however, Johnson employed a more forceful rhetorical response. When receiving help dressing from her personal assistant, Carmen, Johnson noticed Carmen fussing with a pillow positioned between her and her power chair. Carmen explained that she wanted to hide "this furry thing" that Johnson sat on, but Johnson responded firmly, "Leave it . . . Singer knows lots of people eat meat. Now he'll know some crips sit on sheep skin" (213). Although Johnson hesitated to verbalize a discussion with Singer on animal rights, she conveyed a powerful message in an embodied rhetorical response. Through the sheepskin she wore on her body and on her power chair, she issued a response to Singer and what she saw as his impositions on her personhood.

Johnson formed another powerfully embodied rhetorical response to Singer directly through touch at a dinner following their debate when she needed assistance and her aide, Carmen, was across the table. As Johnson describes, "My right elbow slips out from under me. This is awkward. Carmen is opposite me; to make the necessary adjustment, she'd need to stand up, walk around the table, and interrupt the flow of talk. Normally I get whoever is on my right hand to do this sort of thing. Why not now? I gesture to Singer. He leans over and I whisper. 'Grasp this wrist and pull forward one inch, without lifting.' He looks a little surprised but follows my instructions to the letter. He sees now that I can reach my food with my fork. I get the idea that he may now understand what I was saying a minute ago, that most of the assistance disabled people need does not demand medical training" (221). The timing of this assistance was crucial. Johnson used touch to demonstrate and clinch an earlier argument, succeeding in persuading Singer that her care is not necessarily a heavy burden. She *kairotically* grasped hold of a rhetorical situation, improvising and inventing a connection. Rather than bowing to decorum and enlisting Carmen's aid, she made use of the proximity of the person closest to her and used help from Singer to reinforce her argument. In doing this, Johnson used touch as both "immediate sensation and interpersonal affect" (Paterson 153). Touch operated *kairotically* to bolster her argument and to make emotional connections.

When she described the incident to family, friends, and disability rights comrades, however, Johnson experienced resistance from her audience. Simply put, those close to her could not accept the fact that any rhetorical identification may have occurred via this physical assistance. Although she has written the story of her visit to Princeton and told it to many people, she admits that "it proves to be a story that won't settle down. It lacks structure; I'm miles away from a rational argument. The telling keeps getting interrupted by questions" (202). When explaining the dinner incident to a colleague, for example, Johnson struggled particularly with interruptions from her interlocutor. Her colleague was "appalled" that Johnson let Singer "provide even minor physical assistance" at the table and asked how Johnson could put herself "in a relationship with Singer that made him appear so human, even kind" by enlisting his physical assistance (223). When Johnson talked to her sister Beth on the phone about the visit, Beth accused her of "lik[ing] the monster" (225). Although Beth changed the topic, Johnson continued the conversation in her head even after hanging up, defending her position in a self-dialogue. Engaging herself in a question-and-answer session, Johnson finally articulated the implications of her interaction with Singer. Regarding Singer's position on the personhood of people with disabilities, she concludes, "He stirs the pot, brings things out in the open. But my side will win out. We'll make a world that's fit to live in" (226). She values that he articulates difficult views instead of silencing them, and that this articulation invites her response, which she is sure will triumph.

With difficult conversations such as these, Johnson sets out to answer one of her key questions: what does it mean to be a person? Implicitly framing the stakes of this question around the issue of identification, she has found that despite wanting to imagine Singer as a monster or a ghoul, she admits what she calls the "hard part" of the story: "I've come to believe that Singer actually is human, even kind in his way" (223). When looking at him, she experiences what she calls "fellow-feeling" (225). She knows that he "doesn't deserve [her] human sympathy" and cannot form a "logical argument" for why he should (225). Even though she admits that Singer is gaining "a receptive audience" for his ideas, she believes that her point of view will win out (226). Furthermore, Johnson enjoys the "fury that rages when opposing perspectives are let loose" and trusts that "while we struggle, we must also live with our theories and with one another" (228). Johnson calls the heart of her argument against Singer the fact that "the presence or absence of disability doesn't predict quality of life" (205).

Emotions such as sympathy and empathy are also at the heart of Johnson's response to Singer, especially in relation to possibilities for emotional identification with wider audiences. Johnson relates, "If I define Singer's kind of disability prejudice as an ultimate evil, and him as a monster, then I must so define all who believe" similar prejudices (227). She continues, "That would make monsters out of many of the people with whom I move on sidewalks, do business, break bread, swap stories, and share the grunt work of politics. The definition would reach some of my family and most of my nondisabled friends, people who show me enormous kindness and who somehow, sometimes manage to love me through their ignorance" (227–28).

Johnson cannot bring herself to demonize Singer because it would mean cutting herself off from many people in her life, many of whom she considers to be, from a rhetorical perspective, an audience with whom she makes emotional connections and identifications. "I can't refuse the monster-majority basic courtesy, respect, and human sympathy. It's not in my heart to deny every single one of them, categorically, my affection and my love" (228). Johnson holds tightly on to the fellow-feeling among her audiences because she has spent too much time as a "disability-pariah" struggling for kinship, community, and connection and cannot sever any ties she has made (228). After weighing the various forces at work in the situation, she values the emotional reciprocity and the fellow-feeling she experiences through living and speaking with others.

Attempting to settle down her unruly story, Johnson writes, "As a shield from the terrible purity of Singer's vision, I'll look to the corruption that comes from interconnectedness. To justify my hopes that Peter Singer's theoretical world—and its entirely logical extensions—won't become real, I'll invoke the muck and mess and undeniable reality of disabled lives well lived. That's the best I can do right now" (228). Fellow-feeling, muck, and mess form the force of Johnson's response to Singer, one that she acknowledges is hard-won and only temporary. This temporality, based on interconnectedness and lived reality, is *kairotic* and flexible, a response that changes based on the needs of the situation. Johnson is able to register fellow-feeling and partial identification with Singer and others because she critically examines the responses from herself and others elicited by her rhetorical acts of touch, such as her request of physical assistance from Singer at dinner. Yet Johnson establishes limits in the reach of her potential identification. As Cynthia Lewiecki-Wilson notes, Johnson "ends in a kind of stand off, concluding that [Singer's] disability prejudice is not inherently evil though his tenets present a slippery slope that could lead there" ("Ableist Rhetorics, Nevertheless" 74). Johnson, for example, cannot connect with Singer emotionally or intellectually over his argument for animal rights. She writes, "I am still seeking acceptance of my humanity; Singer's call to get past species seems a luxury way beyond my reach" (228). Johnson leaves open the possibility that once her struggle for acceptance is achieved, she may be able to connect with Singer, hoping for identification rather than division.

Disability *Kairos* in Action

As Harriet McBryde Johnson's experience suggests, *kairos* is an action-oriented rhetorical principle, often generating inventive responses to new, unprecedented, or even untenable situations. The tactile elements of *kairos* position it as an especially productive and emotionally generative rhetorical practice for rhetors with disabilities, modeling diverse, nonnormative ways of moving through time and space. In unfamiliar contexts, tactile *kairos* presents a way of moving forward and offers new ways to connect emotionally with audiences.

The journalist John Hockenberry, who uses a wheelchair, demonstrates a sense of tactile *kairos* in shaping productive emotional connections with his audiences, improvising often with the technological tools of his trade to channel emotion. In "Walking with the Kurds," he describes his journey on the back of a donkey across the mountains separating Iraq and Turkey in the aftermath of the first Gulf War. In relating his journey across the mountains with the Kurdish refugee movement he is covering, Hockenberry implicitly draws on transformative notions of disability *kairos* to invite potential identification through proximity. Hockenberry enacts *kairos* improvisationally, using the mode of his body's movements in relation to the other bodies in motion to creatively induce time and space for interaction and emotional connection between himself and the refugees.

Accustomed to using a wheelchair, Hockenberry registered a disruption to his sense of movement in time and space when he traveled atop a donkey for safety and convenience in the rugged mountain terrain. Sitting atop the donkey, he writes, "There were legs below. Stilts of bone and fur picking around mud and easing up the side of a mountain near the Turkish border with Iraq. Two other legs slapped the sides of the donkey at each step like denim-lined saddlebags. They contained my own leg and hip bones, long the passengers of my body's journeys, and for just as long a theme of my mind's wanderings" (22). Hockenberry describes his body at a distance from himself at the outset of his journey. He emphasizes the length of time he has used a wheelchair, also noting how long he has considered his legs to be "passengers" of his wanderings of the mind. In terms of proximity, his own legs are at a distance in this description, once removed from the donkey's legs that provide him with the mobility he needs to cover the refugees' story.

As Hockenberry describes, the familiar experiences of time and space via wheelchair were transformed by his ride on the donkey: "Neither the heroic foot-borne relief efforts, anticipation of the horrors ahead, nor the brilliance of the scenery around me struck home as much as the rhythm of the donkey's forelegs beneath my hips. It was walking, that feeling of groping and climbing and floating on stilts that I had not felt for fifteen years. It was a feeling no wheelchair could convey. I had long ago grown to love my own wheels and their special physical grace, and so this clumsy leg walk was not something I missed until the sensation came rushing back through my body from the shoulders of a donkey" (23). Shared sensation and transference of feeling—the experience of his body in close contact and proximity to that of the donkey—jarred Hockenberry out of his usual rhythms of moving and thinking. Atop the donkey, Hockenberry implicitly revised ancient associations of *kairos* such as Homer's aligning of *kairos* with a potentially lethal space of bodily vulnerability and Hesiod's sense of *kairos* accompanying advice against the overloading of a horse-drawn wagon. *Kairos* became for Hockenberry a strategy for moving with the flow of a story, and disability became potentially advantageous in this scenario as he used his body to move in new ways. Hockenberry grasped a rhetorical opportunity by feeling the weight of himself on the donkey, by stowing away his wheelchair, and by moving another way—he realized that "for the sake of a story, the impulse to toss my own wheelchair to the wind was as natural as carrying a notebook is to other journalists" (26). In this moment he moved flexibly between technologies of assistance, realizing that a notebook, a wheelchair, and even a donkey can be interchangeable.[8] Hockenberry achieved this by adapting himself and his assistive technologies to the situation at hand.

When Hockenberry encountered a gathering of refugees, some of whom were injured, his body's unique positioning in space and time—atop and then dismounted from the donkey—opened him up to connecting productively with his audience. At first, when he was perched atop the donkey, his audience did not know he is disabled. Mehmet, the donkey owner, explained to the refugees that Hockenberry cannot walk, and this information unmasked Hockenberry, who had been previously assumed to be nondisabled. Hockenberry describes a man carrying on his back another man who had a bandage around his waist and was missing a leg. The man looked at Hockenberry, pointed to his wounded friend, and said, "There is danger here. He cannot walk . . . we have here many who cannot walk. . . . Why are you here?" (27).

In response to the man's question, Hockenberry dismounted the donkey, moving closer to the refugees: "I got down off the donkey, sat on the ground, and assembled my tape recorder and microphone. The Kurdish refugees wanted to know why I couldn't walk and if the Iraqis had shot me. Gradually they began to talk" (27). Once they understood the context of Hockenberry's disability, they began to open up. At this juncture Hockenberry also introduced relatively simple but powerful technologies into the rhetorical situation—the voice recorder and microphone. He describes his audience beginning to open up: "'I am a teacher,' said one. 'I am an engineer,' said another" (27). Suddenly "a large man stepped up and grabbed [the] microphone," asking, "What is for American democracy? Bush is speaking of freedom and here we are free? You see us. They send you to us. You, who cannot stand? You are American, what is America now? Why are you here?" (28). Absorbing these questions, Hockenberry reflects, "To him my presence was an unsightly metaphor of America itself: able to arrive but unable to stand. I could not escape his metaphor any more than I could get off that mountain by myself. These were the questions. And so they remain" (28).

When this interviewee suddenly grabbed the microphone, *kairotically* seizing opportunity, Hockenberry experienced a productive blurring of audience and speaker and posed important but unanswerable questions to himself and his country. Reflecting, Hockenberry implicitly drew on the key properties of *kairos* as a means of circulating emotion. "On a donkey among Kurds at the end of a dreadful back-lot surgical abortion of a war, the paths of truth and physical independence seemed to diverge. I had no good answer for the Kurdish man who insisted that there were already too many people who could not walk" (32–33). Although he had no answers, Hockenberry experienced a powerful interaction and transmission of experience and emotion with his audience, as *kairos* circulates among various bodies. Hockenberry, the refugees, and the donkey are bodies in proximity and contact in a transformative rhetorical situation. When a refugee grabbed the microphone, *kairos* was circulated and exchanged among speaker and audience, moving them to new emotional levels.

In turn, Hockenberry, although he did not find easy answers, did find some alternatives to the disability narrative he has been living out for much of his life. In keeping with the "stock figures" of disability that Harriet McBryde Johnson identifies, Hockenberry aligned himself with several "rigid narrative templates," especially his desire to overcome his disability. He writes, "Holding onto that flimsy saddle and feeling each donkey step in my back and in my cramped and throbbing fingers, I could see that my entire existence had become a mission of never saying no to the physical challenges the world presented to a wheelchair" (33). In resistance to this overcoming narrative, Hockenberry realized in Kurdistan "that the world is a much larger place than can be filled by the mission of one man and his wheelchair" (33). This discovery, which changed Hockenberry's emotional relationship to himself, his disability, and his audience, was initiated not only by the feeling of the donkey's movements but also by the feeling he received when he fell off the donkey and had to depend on the refugees and the owner of the donkey to carry him. Hockenberry realized, "If the Kurds had truly left me alone and gone about the business of only saving themselves, I would just have died right there, holding my tape recorder. They did not" (33). With the Kurds, Hockenberry learned the valuable but complicated lessons of interdependence rather than independence. His proximity to the Kurds in body and emotion led to a productive interdependence—Hockenberry told their story and they saved his life. He identified himself with and through them, paving the way for new emotional connections between his immediate audience of the Kurds and his secondary audience of readers.

As in Johnson's case, Hockenberry's identification with his audience was necessarily contingent and therefore partial. When Hockenberry fell and needed to rely on the physical assistance of the Kurds, he began to understand the intricacies of interdependence but still could not answer the difficult questions about his place and his country's involvement in a war. When Johnson's wrist slipped and she needed to enlist the help of Singer, she began to understand Singer with fellow-feeling but could not extend this feeling to all the elements of his argument. Even with these limits on identification, Hockenberry and Johnson showed that new emotional connections are possible when different bodies come into proximal contact with each other in time and space. Not in spite of but because of these partialities of identification, disabled rhetors such as Hockenberry and Johnson establish alternative emotional connections with their audiences.

New Rhetorics Reshaped

Hockenberry, Johnson, and Mairs all achieve opportunities for identification and new emotional connections with their audiences by grasping the *kairotic* opportunities that arise within the proximities of bodies in contact and relation. They revise traditional stereotypes about physical disability and forge emotional connections beyond the one-sided experiences of pity, inspiration, and sympathy, moving toward more transactional emotions with their audiences based on interaction, shared meaning-making, and identification. Touch functions as a crucial vehicle for resisting these limiting emotions and catalyzing new emotional transactions among disabled rhetors and their audiences—disabled, nondisabled, or temporarily able-bodied. Touch changes the relationship among bodies in time and space, *kairotically* bringing bodies in closer proximity and facilitating opportunities for identification.

As Rice suggests, new perspectives on *pathos* produced by efforts to understand the critical elements of affect necessitate an approach to *pathos* beyond emotion that can generate wider-reaching implications. Touch and its attendant affects reach into all the elements of rhetoric, transforming rhetorical situations, the appeals, and other related rhetorical practices. Rhetors such as Harriet McBryde Johnson and Hockenberry certainly illustrate new approaches to the emotions of *pathos* based on the *kairotic* touches and proximities they experience in their rhetorical productions, but they also demonstrate additional approaches to the other appeals of *ethos* and *logos,* via the ways in which touch reshapes their various rhetorical situations and identifications. Johnson, for example, uses touch to craft her *ethos* as much as she does to spur *pathos.* When she decided to shake hands with Singer at the College of Charleston, for example, she did so to bolster her *ethos* as a polite southerner ready to extend some semblance of goodwill and available to engage in civil discourse even with someone whom she believes wants her dead. Once she made this effort, she related that Singer became more human to her; even though she wants to think she caused a visceral reaction in him—flinching—she notes that he seemed ready to deal with her "as a person." The handshake, therefore, reinforced Johnson's *ethos* as it prepared her to register new emotions and judgments about Singer.

Similarly, Hockenberry's use of touch and registering of sensation contribute to his forming *ethos* as much as *pathos.* When he mounted the donkey and reflected on the differences in sensation he felt in relation to his decades using a wheelchair, he opened himself up as a rhetor ready to craft an *ethos* different from his normal sense of character. He realized that he had been living a life in which he unnecessarily individualized his disability, eager to overcome all obstacles by self-reliance and strict independence. Experiencing the walk with the Kurds, however, he realized that a different kind of *ethos* serves him well, and he now understands that "the world is a much larger place" than just one man and his wheelchair. The new emotional connections that Hockenberry makes with his audience hinge on the new ways in which he sees himself as part of an interdependent and interconnected network.

While both Johnson and Hockenberry use a sense of *kairos* based on touch, sensation, and proximity to cultivate new emotional connections to audiences, they also can be seen as using a sense of *mētis* based on touch. Both respond to "shifting terrain" and "ambiguous circumstances," using skills of cunning not based on strength or traditional understandings of ability, and employ tactics that are "multiple and diverse" (Detienne and Vernant 14, 18). They clearly exhibit characteristics of *mētis* such as "resourcefulness," "flair, wisdom, [and] forethought" (3). Their experiences support Marcel Detienne and Jean-Pierre Vernant's suggestion that *kairos* is a kind of *mētis*, as both practices are clearly interconnected.

Both Johnson and Hockenberry bring the interconnections that touch affords to bear on their messages about disability—their *logoi*. Johnson and Hockenberry both invoke the relationships that the experience of disability highlights in order to understand their lives, their places in the world, and their relationships to their audiences. Neither has a direct or specific answer to difficult questions posed to them by their audiences—Johnson cannot answer the student's question about animal rights, and Hockenberry cannot answer the refugee's question about why he was there—but each formulates an alternative response, forming pliable *logoi* rather than logical answers. In Empedoclean fashion, their *logoi* proliferate as their experiences, feelings, thoughts, and emotions multiply and expand through their engagements with others. Johnson describes the "muck and mess" of disabled lives lived well, and Hockenberry relates his lesson that there is more to the world than just one man and his wheelchair. Each emphasizes to audiences that disability is never one thing—it is not a singular experience with independence starring as the main narrative template. Instead disability is many things, a plethora of responses and experiences flexible and responsive to a wide range of situations, audiences, and messages. In this sense the experience of disability is most ripe for experiments in partial identification and productive disidentification.

Most broadly, the interconnections of the appeals of *ethos, pathos,* and *logos* with the practices of *mētis* and *kairos* and Empedoclean *logos* demonstrate the flexibility and porousness of rhetoric. A rhetoric attentive to touch heightens this flexibility, spanning to accommodate the interconnections and overlappings of the elements, situations, and practices of rhetoric. Rhetors such as Johnson, Hockenberry, and a wide range of other people with disabilities frequently use touch to disturb the boundaries of the seemingly tightly compartmentalized elements of rhetoric, trying out new haptic combinations that result in different approaches. Their efforts chronicle the potential for new shapes of rhetoric to come, including those responsive to advancements in technology.

Rhetors with physical disabilities frequently use technologies as vehicles for touch, bringing themselves in closer contact with their audiences. Mairs, Harriet Johnson, and Hockenberry use relatively simple technologies—telephone, recorder, microphone, fork—in *kairotic* ways to grasp rhetorical opportunities. They often exert sophisticated strategies for using these simple technologies. Mairs and Hockenberry, for example, exercise a productive knowledge of both how and when to use technologies in rhetorical situations. For Mairs, this means realizing how being connected to someone over the telephone can inspire writing; for Hockenberry, the knowledge means knowing when to swap his wheelchair for a donkey and when to introduce a microphone and a tape recorder into his interactions with his audience. Johnson knows when to enlist physical assistance strategically for a technology that nearly everyone uses—a fork—to underscore a previous point. Rhetors such as Mairs, Hockenberry, and Johnson practice an art or *technē* of using their devices to the best advantage to facilitate rhetorical opportunity. Instead of just knowing how to use a particular device or assistive technology, they also know when to use it, becoming *kairotic* users. Just as often they invent ways of using technologies in shifting terrains and ambiguous situations, becoming *mētic* users. Contemporary disabled rhetors such as Johnson, Mairs, and Hockenberry, drawing on tactility to combine human ingenuity with technology, model productive strategies from which all rhetors—disabled or not—can benefit, particularly in contexts of new and emerging haptic technologies.

6

Teaching Touch, Touching Technology

Interfaces of Haptics and Disability

In his teachings Empedocles uses an example of a water clock to describe the bodily function of respiration, a description that is also usually accepted as representative of his general theory of pores and effluences. In explaining the composition and function of respiration—a description that also characterizes all other living and nonliving things—he instructs, "All have bloodless tubes of flesh extended over the surface of their bodies; and at the mouths of these the outermost surface of the skin is perforated all over with pores closely packed together, so as to keep in the blood while a free passage is cut for the air to pass through" (Diels and Kranz 31.B.100; Burnet 219). Both air and blood "rush through" these pores and passages in an alternating rhythm. To illustrate this theory, Empedocles employs an example of the use of a water clock by a girl who stops the passage of air and water with her hands. Describing the movement of air and blood in relation to the movement of a hand setting a water clock, Empedocles mobilizes an example centered on manual dexterity and tactile technical ability:

> Just as when a girl, playing with a water-clock of shining brass, puts the orifice of the pipe upon her comely hand, and dips the water-clock into the yielding mass of silvery water—the stream does not then flow into the vessel, but the bulk of the air inside, pressing upon the close-packed perforations, keeps it out till she uncovers the compressed stream; but then air escapes and an equal volume of water runs in,—just in the same way, when water occupies the depths of the brazen vessel and the opening and passage is stopped up by the human hand, the air outside, striving to get in, holds the water back at the gates of the ill-sounding neck, pressing upon its surface, till she lets go with her hand. Then, on the contrary, just in the opposite way to what happened before, the wind rushes in and an equal volume of water runs out to make room. Even so, when the thin blood that surges through the limbs rushes backwards to the interior, straightway the stream of air comes in with a rushing swell; but when the blood runs back the air breathes out again in equal quantity. (Diels and Kranz 31.B.100; Burnet 219)

The girl using the water clock, in Empedocles's description, creates proportions of air and water via her hand, imbuing the keeping of time with a tactile and hands-on approach.[1] In doing so the girl practices a sense of user-generated knowledge in her operation of the water clock, exercising know-how of when to introduce and remove her hand from the flow.[2] The girl who uses the water clock controls the flow of water, modeling a *technē* of touch in which bodies and technologies interact.

By employing the example of the girl and the water clock, Empedocles not only demonstrates his theories but also models an agential, experimental, and tactile method of engagement among users and technologies. The example of the water clock follows a pattern of tactile examples in Empedocles's teachings designed to engage his listeners. From his analogy of the painters mixing paints in their hands, to Hephaestus with his curved hands making proportions of blood and bone, to Aphrodite actively working with her hands, Empedocles depends on the sense of touch to connect to his audiences.[3] Using examples based on touch in his teachings to engage his audiences on an active level, Empedocles imbues his theories with embodied experience.

Pedagogically, Empedocles uses the water clock to illustrate his theory of pores and effluences because it is a device with which anyone listening to him would be familiar: the "clepsydra was a common household contrivance used for transferring small amounts of liquid from one container to another and perhaps for measuring" (Wright 247). Listeners familiar with the water clock can picture themselves using it as they hear Empedocles and probably imagine the feel of the device—shining brass or dry pottery—as it descends into the wet, silvery water. This feel for the device, exerted by the human hand pressing on the surface of water and air, imbues a sense of human agency into all of Empedocles's teachings. The girl with the water clock works with her hands to effect change, controlling the elements, their movements, and their proportions. In this way Empedocles encourages his listeners to engage actively with their world, their technologies, and their bodies by appealing to their sense of touch.

The Interface of Touch

Empedocles's ancient theories, especially his theory of pores and effluences, are applicable to the highly mediated context of the present day, offering ways of instructing contemporary audiences interested in new and emerging technologies such as those based on touch, also known as haptic technologies. As Empedocles's use of devices such as the water clock shows, his theories are directly applicable to machines, even the ones we interact with in the present and imagine for the future. The media theorist Siegfried Zielinski interprets Empedocles's theory of pores and effluences in present-day "media-heuristics" terms, calling it "eminently suitable as a theory of a perfect *interface*" because it includes "double compatibility: size and relative power of the pores and effluences must match so that exchange can take place" and also entails the "reciprocal giving and receiving of attention" (55, 53).[4] Touch is the interface in this reciprocal relationship, as pores and effluences as well as energies and matter match up perfectly and exchange positions through contact.

Thousands of years after Empedocles positioned touch as the basic interface in his theory of pores and effluences, touch continues to function as the ideal and ultimate interface for existing and future technologies such as those based on haptics, a growing area of technological research and development. From the iPad and the iPod to gaming consoles, surgery techniques, navigation systems, and mobile computing devices, haptic technologies situate touch as the "new" and ideal interface through which to connect body and machine. As a general concept, the notion of an interface is used to describe a wide range of bodies, materials, and entities that come in physical contact. Both Empedocles's and Aristotle's theories on touch are relevant to current haptic-interface technology discourses and practices, especially in relation to the varieties of embodied experience that touch affords. Empedocles's and Aristotle's theories offer avenues for critical inquiry and interrogation of touch as a purportedly ideal, accessible, and natural interface for bodies and technologies. While Empedocles and Aristotle may differ on their specifics of their theories of touch and how human bodies connect to the material world, they are both interested in exploring a shared concept—the theorization and materiality of the interface—in ways that inform contemporary discourse and practices surrounding haptic technologies.

It is exactly, for example, the perfection of the tactile interface in Empedocles's theories that troubles Aristotle, leading him to consider how bodily difference affects touch. In his catalog of the senses, Aristotle notes that it is challenging to identify the "underlying unity of touch" (*De Anima* 422b). Unlike sight, which corresponds to the eye, and hearing, which corresponds to the ear, touch has no single organ and instead functions via the medium of the flesh. As Aristotle comments, "The tangible differs from the visible and the audible; for we perceive the latter in that the medium itself produces some effect in us; whereas the tangible object does not affect us *through* the medium so much as *with* the medium, simultaneously, as when one is struck on a shield. For the shield does not strike its holder *after* it is itself struck; but the two are struck at once" (423a). Therefore, in Aristotle's view, there is no "perfect" relationship between touch and the world, since "flesh, then, is the medium of touch" and "tangible objects vary therefore with differences of body as such" (423a–423b). By illustrating the complexity of the medium of flesh with the example of a person carrying a shield, Aristotle emphasizes the importance of the medium. As Ronald Polansky explains, "The directness of contact in spite of the possible mediation of shield is because the principal medium of touch . . . is the flesh that is part of the body rather than the outside expanses of air or water that serve for media of the distancing senses" (329). Flesh, then, is a *medium* of touch for Aristotle that possesses the potential for generating affective response, according to bodily difference.

In contrast, in contemporary haptic technologies touch is often rendered as a natural expression that dissolves the medium or interface and neglects bodily difference. Multitouch technology is frequently described as a technology in which "the interface just disappears" ("Jeff Han"). The commercials for Apple's iPad position the haptic interface as something that is simply "magical" ("What Is iPad?"). With close-up shots of able-looking hands on various screens performing various tasks—flipping pages in a book, thumbing through photos, playing a keyboard—Apple implies that manipulating a haptic interface on a screen unites body and technology seamlessly. Mark Paterson, in his study of haptics, questions the apparent seamlessness of haptics, asking, "If in reaching out, touching, or manipulating an object we register such tactile contact through the medium of flesh, whence comes the illusion that there is no intermediary?" (158). He notes that in contrast to Aristotle's emphasis on the medium of flesh, which is "extendable through prosthetic means," in contemporary haptic practices "our contact with things is erroneously perceived as direct, as unmediated" (17).[5] The interface or medium disappears even as it is utilized.

From a disability studies perspective, Aristotle's attention to the medium of flesh is significant, as it points to ways in which touch invites considerations of proximity and immediacy between bodies and technologies from the perspective of vulnerability. As a sense driven by proximity, touch leads Aristotle to consider the complex ways touch affects lived experience. In fact Edith Wyschogrod posits that Aristotle finds the element of proximity in touch so "vexing" that he disrupts his established model of the senses, adding a consideration of "vulnerability" to the medium of touch (194). Wyschogrod argues that "flesh is lived as vulnerability in every tactile encounter while the perception of tactile qualities, hardness, roughness, etc., is ancillary to this lived vulnerability and follows from it" (194). The sense of touch calls attention to vulnerable bodies in that bodies are vulnerable through and by touch.

Aristotelian and Empedoclean theories of touch, when applied to contemporary haptic interfaces, provide a context for critically examining ableist assumptions that underlie the expected relationship between bodies and technologies. Aristotle's attention to vulnerability, for example, draws attention to the fact that all of the various pairs of hands in the popular iPad commercials are perfectly formed and normatively shaped, a metonymy for the able-bodied, non–visually impaired user implied in the advertisements.[6] Empedocles's theories on touch and technology draw attention to the act of touch itself, reminding audiences that touch is not simply a magical and invisible action, as portrayed on iPad commercials, but that user knowledge and practices can be agential sources of meaning-making. Critical exploration of the sense of touch in haptics from perspectives such as these is necessary to understand more comprehensively how touch functions as an interface, especially what kinds of bodies are assumed as ideal for haptic interaction and what kinds of bodies are denied. Assumptions about ability and disability—what kinds of bodies can and should use haptics—shape the design, practices, and discourses surrounding haptic interfaces.

Investigating issues involving bodies, technologies, and touch is crucial in the developing area of haptic technologies because this investigation contributes to what Bruno Latour has called the "unblackboxing" of technology. Latour writes that the word "black box is used by cyberneticians whenever a piece of machinery or a set of commands is too complex. In its place they draw a little box about which they need to know nothing but its input and output" (2–3). Technologies or codes, then, are understood as ready-made, rather than in the making—"no matter how controversial their history, how complex their inner workings, how large their commercial or academic networks that hold them in place" (3). Opening black boxes means questioning the input and output of accepted practices and investigating the discourses of existing and developing technologies. Applying Latour's method to haptic technologies means critically questioning the interface of haptic technologies, rather than simply accepting touch as the "input" with a predetermined "output."

As I explore in this chapter, "unblackboxing" haptic technologies requires asking questions about touch in relation to accessibility, ability, and disability. Both Empedocles and Aristotle raise key questions and complexities of touch that still inform contemporary interactions with interfaces and technologies. At the most basic level, both Empedocles's and Aristotle's theories ask questions about what constitutes the sense of touch as an interface or medium between the body and the world. The questions surrounding Empedocles's theory of the perfect interface and Aristotle's considerations of bodily vulnerability are relevant not only to the design and accessibility of contemporary technologies but also to how we talk, write, use, and teach technology in the classroom.

In the following sections I consider the touch interfaces involved in haptic technologies, which have become ubiquitous with the success of Apple products such as iPod, iPhone, and iPad. Specifically, I explore how to teach students in the writing and technical writing classroom to explore the black box of haptics and to be more critical of the histories, discourses, and practices that accompany the use of haptics. After exploring how the history of haptics is bound up with discourses of ability and disability, in the study I detail, students learned to apply this critical literacy to the study of accessibility in haptic technologies. Together with students I interrogated the assumptions that often accompany haptics and their use, including the portrayal of haptics as interfaceless technologies, seamlessly integrated with the almost exclusively able body. After exploring my methods and perspectives for this study, I describe how I led students to explore touch and their status as agential users of haptic technology more critically.

Methods and Perspectives

I combined teacher-researcher and disability studies methods in my classroom study of touch and haptics. I consider myself a teacher-researcher, or someone conducting "systematic, intentional inquiry" while teaching in a classroom (Cochran-Smith and Lytle 23).[7] Crucial to my classroom research is the disability rights movement's mantra, "nothing about us without us," which demands that people with disabilities speak for themselves rather than be spoken for in everything from research agendas and policy making to discourses of representation. In terms of research methodologies, this activist mantra means that people with disabilities become the subjects and not the objects of research and often collaborate with allies. I saw my class as participating in this larger effort, and I actively attempted to reach students with disabilities registering for classes by posting to listservs and in the university disability services offices. When writing the course description that students would reference to schedule courses, I described my writing course as paying particular attention to concepts of disability, technology, and accessibility in addition to the more typical concepts of audience, context, and argument. I sought out an accessible classroom—one that could be reached by elevator for students with physical disabilities and conducive to multimodal pedagogies through technology. At the start of the class six students identified as having disabilities and two students identified as having had serious illnesses in the past, and during the course two students experienced temporary disabilities. Nearly every student in the class wrote about a close family member, friend, or other relation with a disability.[8]

I relied heavily on my subjective and self-reflective observations of the class, particularly concerning class discussions. To balance these observations, I also depended on students' reflections of key concepts covered in these discussions. Toward this end I integrated regular responses from students through online journaling on a collaborative wiki, a class Web site that enabled users to add and update content. Students regularly responded to questions I posed about the intersection of disability, accessibility, and disability studies in regular in-class writing exercises and on the wiki. Students also regularly reflected on their assignments, particularly in a larger wiki response at the end of the semester. Classroom discussions and student assignments formed the bulk of my data. How students responded to, struggled with, or succeeded with discussions and assignments were important facets of my study. The study underwent Institutional Review Board examination and included a process of informed consent.[9] During the first weeks of class, I described the study to the class multiple times and made students aware that their participation was voluntary.

The multimediated writing classroom is the ideal space in which to retheorize haptic technologies and to test their accessibility for people with wide ranges of ability. To familiarize students with disability studies perspectives, I assigned excerpts from Simi Linton's *Claiming Disability,* particularly the chapter "Reassigning Meaning." Linton's argument focuses on reassigning the negative meanings usually associated with disability as deficit. Like other disability studies scholars, such as Lennard Davis and Rosemarie Garland-Thomson, Linton understands ability and disability as part of the same system: the "absolute categories *normal* and *abnormal* depend on each other for their existence and depend on the maintenance of the opposition for their meaning" (23). Linton uses terms such as "nondisabled" to center disability and place the category of "nondisabled" in "the peripheral position in order to look at the world from the inside out" (13). To this end I modified a classroom activity described by James C. Wilson, in which students "construct their own norm(s) and . . . design a 'normal' human," an exercise that produces divergent responses that disrupt students' typical classifications as they realize that "norms are cultural constructions that depend on particular frames of reference" ("Making Disability Visible" 153–54). I extended Wilson's exercise to the specific application of haptics by asking students to design their own ideal or universal types of haptic technologies. Students quickly found that the ideal for one person differs significantly from the ideal for another. Different students designed different technologies based on individual experiences of embodiment, finding it difficult to come to any agreement on what constitutes a "universal" or even "normal" user.

This exercise also resisted limiting simulations of disability in classrooms, which often ask students to "try on a disability" for a day. Jim Swan, drawing from an online discussion on the Disability Studies in the Humanities (DS-HUM) list, asks "whether classroom simulations of disability serve as effective teaching tools or just confirm stereotypes and prejudices" ("Disabilities, Bodies, Voices" 287–88). He explains that a "repeated criticism was that simply trying on a disability (e.g., spending a day in a wheelchair) cannot convey the continuous, no-time-out experience of being impaired. Worse, it focuses too much attention on the impairment and too little on the social construction that turns impairment into disability. The risk is that a simulation will simply confirm nondisabled persons in their belief that they are normal" (288). An alternative is "to consider how the technological culture, the materially constructed environment, which is the result of specific design and policy decisions, acts to enable even those who think of themselves as already enabled on their own" (288). I extended this alternative to the classroom by inviting students to explore the technological culture of haptics, including how these devices are materially constructed to enable or disable users. The cultural discourses that accompanied these technologies also informed our discussions, especially about disability and accessibility. Following Linton, this examination of haptics aimed to draw on "the perspective and expertise" of people with disabilities that have so often been "silenced" (13). In particular, I guided students toward noticing and studying the productive ways in which people with disabilities have engaged with the sense of touch as an interface in their daily lives, both with and without technology. The aim of this instruction was to encourage students to notice more critical and attentive ways of interacting with touch and to apply these strategies to their own use of touch in their everyday lives, especially in relation to haptic technologies. The primary objective was to show students that the sense of touch is not simply a neutral or natural interface that seamlessly connects bodies and technologies but is overlaid with assumptions about bodies, minds, and interactions.

Situating Haptics: Now and Then

To begin a critical investigation of haptics, I guided students through an initial discussion of haptics in their everyday lives. In an early class discussion students and I brainstormed a list of technologies that integrate haptics and identified their features, including the most salient ones involved with their use. One goal in this early discussion was simply to make students notice the haptic technologies with which they interact on a daily basis or are exposed to in culture. Students tend to focus on audio-visual features of technologies, but this discussion encouraged them to pay attention to how they often use the sense of touch to interact with technologies. I wanted to encourage students to begin to notice and then analyze how they interact with haptics and the cultural discourses and assumptions at work in advertisements for haptics, similarly to how they may analyze a visual or textual representation in the classroom. Students identified products such as iPod, iPad, iPhone, Droid, htc/T-Mobile My Touch, and gaming systems such as Wii and Xbox Kinect, among other technologies. I asked them to describe how these technologies are marketed, including the advertisements and branding strategies they noticed about particular products. I specified this question by asking them to identify how the ways these technologies are advertised and marketed contribute to the construction of cultural discourses, assumptions, and logics about haptic technologies. In particular I asked them to consider the relationship of embodiment to these discourses.

In wide-ranging small-group and full-class discussions, students identified several cultural logics of haptics. One group of students shared what they called the "slick aesthetic" of products such as Apple. Advertisements for these products emphasize how easy it is to navigate and use the products via touch: a simple flick of the wheel of the iPod, the completely touchable faces of the iPhone and the iPad, and the ease of wearing a product such as the iPod Shuffle by snapping it to clothing, a purse, or a backpack strap. Students characterized these technologies as meant "to work with" the body, moving with us effortlessly and keeping us connected to each other and to media in a range of environments. Students described the appeal of technologies that are light, handheld, and easy to transport, noting that this promotes a sense of "harmony" or seamlessness between body and technology. As one student described, the Apple commercials such as the one called "What Is iPad?" frequently show no faces and simply many images of seemingly anonymous hands and phones, ostensibly reinforcing "the closeness between user and phone" and "intensifying the feeling of need" for the phone. In affective registers students noted physical attachments to their technologies, feeling "naked" or "anxious" if they left phones at home or were without laptops an extended amount of time. One student described a feeling of "helplessness, loss and disconnection" if one misplaced a phone.

Another group noted what they called a "creepy yet alluring element" of other haptic technologies. Referring particularly to a series of commercials for the Droid mobile devices and operating systems, students described feeling "weird" and "uncomfortable" with the suggestion that they "become their technologies" by using them. In the commercial a human user's hands, pecking frenetically at the device, morph into spindly, androidlike extensions ("Motorola Droid 2"). Close-up shots in the commercial show human skin being transformed into a metallic casing or hardware similar to an insect exoskeleton. Students in this group described feeling that this image was "too close for comfort" and were "unsettled" by the suggestion that their bodies will become technological. Extending this idea, they described feeling that technology sometimes encroaches on their everyday lives, demanding that they be constantly present, in touch, and reachable. Students noted regret at not being able to have dinner without constantly checking their phones, and yet they stopped short of wanting to "cut themselves off" from their phones completely, even for a short time. Several students in this group suggested that this may be an indication of technological addiction.

In an effort toward working through these different views of technology and embodiment and investigating their development, I guided students through a critical history of haptics, asking them to pay special attention to how the sense of touch has been marshaled to suggest certain kinds of relationships between bodies and technologies. Throughout this exploration I reminded them to ask questions about how touch functions to construct bodies in relation to technologies in both productive and limiting ways, suggesting that the cultural discourses, logics, and assumptions of new haptic technologies often possess roots in historical approaches to technology. My primary goal in this exercise was to begin to inspire students to create a set of critical literacies for engaging with haptics historically and in their contemporary everyday lives.

We started by exploring the beginnings of a history of virtual and haptic technologies, focusing on how touch functions in the quest for the perfect interface between body, mind, and technology.[10] We surveyed the objectives of research conducted by MIT professor Vannevar Bush in the 1930s and 1940s and the work of Edwin Link, who patented the Link trainer, a flight simulator. As the digital media theorists Benjamin Woolley and Ken Hillis describe, Link's flight simulation utilized touch as a proxy to simulate the sensations of force and resistance through physical contact with a joystick in a mock cockpit. Simulation hinged on the sense of touch in these experiments, in an attempt to dissolve not only the interface between body and technology but also the line between reality and virtuality.

Continuing our focus on the role of touch in the historical search for the ideal interface, we also investigated the aim of military advancements in force feedback and virtual simulation development in the early 1960s, which led to computer scientist Ivan Sutherland's development of the device called Sketchpad. This interactive program enabled a user to hold a light pen to create designs on a screen to be stored, retrieved, and superimposed (Sutherland "Sketchpad"). As Hillis describes, "In a sense, the light pen (the precursor of the computer mouse) guided the human hand into a conceptual integration with the computer technology" (12). This conceptual integration, designed for the human hand, exercised the desire for what Sutherland calls the "ultimate display." Touch functioned crucially in the development of this ultimate display, connecting body and machine through an interface. As the theorist Myron Krueger relates, "The ultimate interface between the computer and people would be to the human body and human sense" (19). Sutherland's desire for the ultimate interface questioned the basic relationship between body and technology. In his description of "The Ultimate Display," Sutherland speculates that a "display connected to a digital computer . . . is a looking glass into a mathematical wonderland. . . . There is no reason why the objects displayed . . . have to follow the ordinary rules of physical reality. . . . With appropriate programming such a display could literally be the Wonderland in which Alice walked" (506–8). Sutherland's device, a head-mounted three-dimensional display, attempted to reach this ideal of the ultimate display—one that fully immersed the user and vanished the interface between the body and the technology. Because of the weight of the display's attendant technologies, however, it had to be suspended from above the user, demonstrating the physical limits with which designers continued to contend.

As students and I explored the early research and development of virtual and haptic technologies, we paid particular attention to how touch functions both as the ultimate interface between humans and machines and as the interface that researchers ultimately desire to dissolve. The tension between these two desires plays out most obviously in the sense of touch, but it also operates more subtly in questions about rationality and cognition. Describing the Sketchpad, the computer theorist Theodor Nelson illuminates these tensions: "You could draw a picture on the screen with the lightpen—and then file the picture away in the . . . memory . . . magnify and shrink the picture to a spectacular degree. . . . Sketchpad . . . allowed room for human vagueness and judgment. . . . You could rearrange till you got what you wanted . . . a new way of working and seeing was possible. The techniques of the computer screen are general and applicable to everything—but only if you can adapt your mind to thinking in terms of computer screens" (quoted in Rheingold 91). Like Sutherland, Nelson considers the limits imposed on thought by the constraints of technology. Adapting one's mind to thinking in terms of the screen may mean restricting creativity and technological advancement. By questioning the rationalist limits imposed by a technology that purports to reduce the boundary between body, mind, and technology, students and I considered how Sutherland and Nelson may also be asking what cost dissolving the interface exacts.

Students and I explored technological advances dating from World War II, such as the development of the Electronic Numerical Integrator and Computer (ENIAC), a device for automating predictions of missile and bomb trajectories. As Hillis describes, "In 1944, researchers at MIT's Servomechanisms Lab, using digital equipment akin to ENIAC, successfully demonstrated that a light-sensitive, handheld detector wand, when pointed at a television-like screen adapted from radar technology, could select or 'highlight' individual dots pre-programmed to move like bouncing balls across its surface," an action that "bore similarities to reaching out to touch or contact an object" (4). Touch, in the ENIAC device, occupied the crucial boundary between body and technology and became instrumental in the subsequent development of universal computers and technologies. ENIAC technology is credited as exemplifying the type of technology that Alan Turing would soon identify in his concept of the universal machine (Hillis 3). According to Turing, a universal machine is a machine that, once rendered, would make it "unnecessary to design various new machines to do various computing processes" (441). In other words, a universal machine is the machine of machines, something that today's ultimate personal computer still aspires to be. The sense of touch is crucial to this universality because it draws on the basics of embodiment—particularly the sense of touch in the body's relationship to the computer.

Turing's universal machine can be understood to universalize the user as much as the machine. Hillis describes Turing's machine as "a step in conceptualizing the electromechanical simulation of our selves" and critiques the machine's attempt at "timeless universality" by suggesting "an irreducible element" of the self (11). The media theorist Jay Bolter, describing the Turing machine, sums up this link to self and machine: "By making a machine think as a man, man recreates himself, defines himself as a machine" (13). Bush's well-known and prescient 1946 discussion of the personal computer raises similar questions regarding the limits of a universal approach. Bush, in "Memex Instead of Index," urges:

> Consider a future device for individual use, which is a sort of mechanized private file and library. It needs a name, and, to coin one at random, "memex" will do. A memex is a device in which an individual stores all his books, records, and communications, and which is mechanized so that it may be consulted with exceeding speed and flexibility. It is an enlarged intimate supplement to his memory. . . . In the outside world, all forms of intelligence, whether of sound or sight, have been reduced to the form of varying currents in an electric current in order that they may be transmitted. Inside the human frame exactly the same sorts of processes occur. Must we always transform to mechanical movements in order to proceed from one electrical phenomenon to the other? (32).

Hallmarks of ability such as speed, flexibility, and memory are valued in this ideal machine. As a consequence, the "self" that this universal machine defines is one that is not only nondisabled but hyperable. Yet, like Sutherland and Nelson, Bush questions if "the human frame" and the machine must mirror each other. In particular he questions whether the human form—one implicitly able-bodied—must transform to fit the computer, which is fast, flexible, and in possession of a prodigious memory.

In a group discussion students, adding to their previous conversation about the appeal of the "slick aesthetic" of technologies, identified this tension suggested by Bush as the "our bodies our machines" effect, explaining that we desire technologies that link up effortlessly with our bodies and our daily activities. We imagine technologies that resist physical limitations, possessing limitless battery life, wireless capabilities, and endless options; we desire our machines to operate as "superhumans." Increased hallmarks of ability—limitless speed, flexibility, and memory—mark the desire for the ultimate interface and the universal user.

In class discussions I raised the possibility that this tension can be more comprehensively understood by interrogating assumptions about the abilities of this typical human mind and user imagined in these cultural discourses and logics. I asked students to draw connections between the quest for a universal machine and perfect interface and the ideology of ableism that underlies these desires. Does the desire for a universal machine create a universal user that can include people of diverse abilities and disabilities or is the preferred universal user nondisabled and neurotypical? There is no easy answer to this question, but thinking through it led students to note the desire for a universal machine as related to the assumption of a universal user—one that is imagined in current cultural advertisements, images, and practices as able-bodied, neurotypical, and free from any difference. Most crucially, working through the questions that surfaced in our short and select overview of the history of haptics gave the class an opportunity to investigate the sense of touch in a critical context and prepared them for intervening in the present-day cultural discourses of haptics.

As we completed our partial study of the history of haptics and virtual technologies, I guided students toward noticing three current cultural discourses rooted in the early research and development of haptics that continue to circulate today. Current cultural discourses of haptics position touch as the ideal interface between human and machine, often with the assumption that touch makes the interface disappear and the connection between body and technology seamless. Touch remains untheorized and uncritically accepted in this assumption. A related assumption in advertisements and marketing of haptics is that predominantly universal and able bodies are the users of haptics. These two assumptions coalesce in a final assumption that affects all users of haptics in which the cultural discourses and advertisements often surrounding haptics discourage users' knowledge by assuming that users already know how to use them or cannot add anything to them. Summing up our study of the history of haptics in these three contemporary assumptions prepared the class to explore how the study of ability and disability necessarily intersects with and even potentially revises each of these pervasive discourses.

Haptics Today

The desire for the ultimate interface continues in present-day haptics research and development. As in discourses of virtuality, touch in haptics functions both as the ultimate interface and as an invisible interface that researchers and developers seek to dissolve. Steve Jobs of Apple, in unveiling the iPad, which fully integrates touch technology, described the device simply as "magical," a descriptor repeated in commercials (Jobs; "What Is iPad?"). Jefferson Han, a developer of multitouch technology, calls haptics an "interface [that] just disappears" ("Jeff Han"). Students and I interrogated these claims, situating them as extensions of the historical search for the ideal tactile interface between bodies and technologies.

We viewed Han's 2006 Technology, Entertainment, Design (TED) demonstration of multitouch technology, which was well received by its live audience and quickly became popular online in video form on YouTube and other spaces. Running his hands across a modified screen the approximate size and shape of a drafting table, Han manipulates the borders of pictures, draws lines, pulls up a floating keyboard, and even transfers his own body heat onto the screen, all by simply touching the screen rather than using a mouse, stylus, or keyboard. Although Han notes that the technology of multitouch "isn't that new," what is new is that it is low cost, extremely scalable, with high resolution and "newfound accessibility" ("Jeff Han"). These attributes, according to Han, make multitouch more "universal" and "completely intuitive." Implying a natural seamlessness between the body and technology, he remarks that there is "no user manual" for this technology, no desktop and no mouse. The demonstration also attempts to persuade the audience that touch technology offers users an "ultimate" experience. Students readily drew connections between Han's description to Sutherland's "ultimate display" and Turing's "ultimate machine." Emphasizing an interface that "disappears," Han says that the technology changes "how we interact with machines." Rather than making users conform to the interface, Han posits that "interfaces should conform to us." Multitouch technology operates on this assumption, emphasizing no need for a manual or mouse. Instead of a technological interface, the interface becomes the body itself, as users simply connect physically with the screen. In fact multitouch technology, with its potential to accommodate more than one user, suggests that it ventures beyond the body-machine interface and toward the seamless collaboration of bodies and machines.

I drew students' attention to different perspectives. Despite claims that haptic technology reduces the interface between body and technology to near imperceptibility and is accessible to a wide range of users, multitouch technology excludes some of the users for whom it was originally designed. As Bill Buxton has shown, people with low vision and blind users are not accommodated by many haptic technologies ("Multitouch Systems"). This inaccessibility is particularly problematic because many of the research advancements that have made haptics available to the general public are built on the foundations of research for people with vision impairments and other disabilities. Noting these limitations, designers such as Ben Challis and Alistair Edwards have developed principles for haptic research that include blind users, but few large-scale technologies accommodate them comprehensively ("Design Principles for Tactile Interaction"). Apple's Voiceover, a screen reader for the iPad, is a built-in accessibility feature for blind and low-vision users, but some users feel that one must have a "strong spatial sense," including a sense of knowing where icons reside in relation to each other on the surface of the iPad, in order to use it effectively (Ziegler "Blind iPad Review"). In addition, although developers of haptics claim that they are "intuitive" and so user-friendly that no manual is needed, what is intuitive to the neurotypical, nondisabled, psychologically "normal" individual differs from what is intuitive for a person with a physical, cognitive, or psychological disability.

By paying attention to how haptics are advertised and characterized in popular cultures—from TED demonstrations to Apple commercials—students became tuned to the underlying assumptions of contemporary discourses of haptics that claim to be interfaceless, understanding that in fact this interface requires new interrogation, critical analysis, and sustained attention. Rather than having students understand haptics as "magic" or touch as "interfaceless," I encouraged them to find ways to pay more attention to touch and to ask questions about the kinds of bodies and minds that discourses and practices of haptics both reflect and shape.

Teaching Touch

To interrogate the assumption that touch simply dissolves the interface between bodies and technologies, I guided students toward defining touch as an interface by noticing the meanings it conveys, particularly in contexts of disability. Janssen pharmaceutical company, for example, has developed a 3-D video simulation called "Mindstorm," described as a representation of what it "feels" like to have schizophrenia. A voice-over in the video simulation informs the audience that "schizophrenia can provide powerful experiences" and that what they will "experience" is a "severe episode of auditory and visual hallucinations" ("Mindstorm"). The voice continues, explaining that "the symptoms you will experience represent a compilation of a range of sensory experiences, as reported by actual patients," including "sights, sounds, wind or even scents." The simulation takes the viewer through several typical everyday activities—getting up in the morning, washing up, drinking coffee, ordering pizza—all from a first-person perspective of someone with schizophrenia who, while doing these activities, also hears voices, hallucinates, and feels paranoid. A particularly powerful experience in the simulation occurs at the end, when the patient, whose face is never revealed in order to heighten the effect of looking out from his eyes, throws down his bottle of medication, succumbing to the voices in his head saying it is poison. A female caretaker pleads with the patient to take his medication if he wants to get better. After showing this video in class and reading descriptions of it in popular media, I asked students what identifications or disidentifications are encouraged and discouraged through this simulation. What assumptions do virtual reality and 3-D simulations carry with them in this simulation of the experience of what it feels like to have mental illness? I asked students to discuss in small groups the omission of touch from the simulation description. I then guided them to consider the purpose, audience, and context of the simulation in addition to identifying the appeals of *ethos, pathos,* and *logos* it mobilizes.

Although "Mindstorm" employs 3-D simulation and virtual reality, technologies based on the sense of touch, the narration preceding the simulation emphasizes other senses, such as sight, sound, and smell. Students noticed that the narrator alludes to the sense of touch in the description of the feel of the wind that the audience will experience, but they reported that touch is so naturalized in the larger description of the virtual experience that it barely receives comment. Touch barely receives mention even in designers' and consultants' descriptions. Only in his concluding comments about the "virtual hallucination experience," for example, does the consultant Gahan Padina mention touch: "In psychology and psychiatry both, we've attempted . . . to try and simulate the feelings and the experiences that patients have, but, I think, with new technologies, in particular, being able to project in three dimensions, using the different modalities—smell, sensation and touch—makes it a much more powerful experience" (Lee "Sympathy through Technology"). The power of this experience is based on a range of sensory perceptions, but touch is one sense that remains unexplored and uncritiqued.

In a class discussion I connected this omission to a pervasive contradictory cultural logic in which virtual reality seeks to simulate the world but also to deny the interface through which this simulation occurs. The media theorists Bolter and Richard Grusin call virtual reality "the clearest (most transparent!) example of the logic of transparent immediacy," a logic positing virtual reality as "immersive, which means that it is a medium whose purpose is to disappear" (161, 21). Virtual reality is known as the "cultural metaphor for the ideal of perfect mediation" (161) because, like haptics, it desires to dissolve the medium itself. The logic of transparency functions in haptic technologies and virtual reality as a "perfect interface" because it is an interface that ideally disappears.

In calling attention to touch in the "Mindstorm" simulation, I led students to intervene in emerging discourses of technology and culture that attempt to treat the sense of touch, especially among bodies and technologies, as a naturalized, seamless interface. Rather than naturalizing touch, students learned to explore its meanings and limits in the "Mindstorm" simulation. Rather than accepting uncritically that the simulation is what it "feels" like to have schizophrenia, students were more likely to ask crucial questions about why a certain technology is used to simulate a psychological condition. Touch is the interface that enables this simulation, but students also called attention to other forces at work. They noted, for example, that Janssen is a pharmaceutical company with vested interests in selling their product.

The interests of Janssen, a drug manufacturer, reside in using the virtual reality simulation to persuade people with schizophrenia and their caregivers to make sure their medications get taken, an action with the objective of returning schizophrenics to a more normal psychological state, without sensory hallucinations and irrational thoughts. Janssen leaves relatively untroubled both the desire to simulate schizophrenia and for users to experience it. Their strongest appeal is a logical one in that they seek to convince their audience—people who do not have schizophrenia themselves but have someone close to them who does—that the right pharmaceuticals can make the world make sense. Students noted that the audience never sees the patient's face and is therefore put in the perspective of the person with schizophrenia, the camera angle looking out from their eyes, ostensibly increasing identification. As a class they expressed skepticism that there is a clear way to make sense of the experience of schizophrenia and invited possibilities for understanding from the perspective of a person with schizophrenia.

Instead of simply accepting that virtual reality and haptic technologies provide an accurate simulation of what it feels like to have schizophrenia, I directed students to question the desire to use touch to simulate this experience and connect it to the larger cultural desire in the history of technology to develop the perfect interface. To a certain extent, this questioning aligned with the alternatives to classroom disability simulation exercises that Swan, Wilson, and others have explored, in that students learned to focus on the underlying assumptions that drive simulations rather than the simulations themselves. Rather than accepting haptics as "interfaceless" or "magic" technologies, students understood them as technologies situated in a specific material and technological culture, with attendant discourses that need to be critiqued and examined. The decision to take a medication, for example, cannot be made easily by a few minutes' worth of a simulated schizophrenia experience; instead it takes careful consideration of side effects, a person's specific situation, his or her past experience with medication, and any other mediating factors.

Touching Technology

Extrapolating from our study of the history of haptics and virtual realities, I led students toward exploring how current haptic technologies also encourage a use and understanding of the sense of touch that relies on assumptions of universality. Many designs and discourses of haptics assume that all users interact with touch and technology in the same way. The much-anticipated release of Apple's iPad, a device poised to popularize haptics in unprecedented ways, produced reviews and demonstrations of the product that emphasized the universality of its interface. Videos, reviews, and demonstrations of iPad often focused uncritically on the haptic experience. Xeni Jardin, a reviewer of the iPad, called it a "touch of genius" and wrote, "It strikes you when you first touch an iPad. The form just feels good, not too lightweight or heavy, nor too thin or thick. It's sensual. It's tactile." This "feel good" and "just right" interface projects a desire to embrace a wide range of users. Within only a few days of the release, the tacit assumption was that anyone could use the iPad just by picking it up and touching it. As noted, commercials almost exclusively implied an able-bodied user of the devise—perfectly formed, smooth-moving, and sure hands populated advertisements. Videos of toddlers and even cats and dogs using the iPad surfaced and were widely viewed; a popular human interest story focused on a ninety-nine-year-old woman who had never used a computer but who instantly understood the haptic interface (Parr).[11] The cultural assumption is that anyone can use a technology that operates via the sense of touch and that anyone with the sense of touch—basically anyone in general—is a universal user. In haptic technologies the sense of touch is used to construct a universal user who simply mobilizes his or her body to interact with a device.

To resist the assumption that everyone interacts the same way with technologies that integrate touch, I challenged students to find examples of people with disabilities who resist this assumption. I drew students' attention to blind users' reviews of the iPad, challenging them to explore the spectrum of diverse ways in which a user can use touch to interact with technology. We read Matilda Ziegler's "Blind iPad Review," which lauds certain accessibility features of the iPad but also contains suggestions about increasing functionality for users with blindness. Ziegler offers the following proposition, speculating that "if a blind user wanted to use a certain app a lot, such as the typing virtual keypad feature, a tracing could be made of the positions of the icons and someone could cut out an overlay of light plastic, like a glorified check writing guide or a stencil. That way, a blind user could tactilely locate the positions quicker. Imagine a sheath of light plastic overlay cut-outs the shape of the screen for different standard uses, such as typing or web surfing, carried in a sleeve or pocket inside the front cover of the iPad case." Ziegler's suggestion of a light plastic stencil to overlay on the iPad not only resists the assumption that all users approach haptics the same way but also reasserts the presence of touch with the proposal of an additional plastic interface between body and machine.

People with cognitive differences also resist the assumption of the universal haptic user. Recent technological developments for students with autism designed for use on iPads, such as "Apps for Autism," show how some students with autism thrive with a haptic-based interface, especially in the areas of communication. On the television news show *60 Minutes,* a segment on the apps featured the testimony of teachers and parents using iPads to work with students with autism who experience difficulties speaking and communicating, integrating video footage of a range of students with autism interacting with iPads ("Apps for Autism"). In this video autistic students use their frequently wavering hands to touch and swipe at a screen, often while rocking and exhibiting nonnormative motor control. Sometimes teachers or aides extend and steady their arms while they touch the screen—in a way similar to the physical assistance provided in facilitated communication. While the iPad is not necessarily a "magical" device for "unlocking" the child inside, as some of the rhetoric of the segment suggests, and experiences with the apps differ by student, the video evidence shows that the haptic user need not be physically or cognitively able-bodied or "normal" in order to use touch to interact with technology. In resistance to the historical tendency of haptic research, design, and development to desire a universal, able user, students with autism demonstrate different possibilities. Experiences and the expertise of people with autism in the designing and implementation of apps for autism are crucial for future user-centered research and development to ensure that nonneurotypical ways of moving, processing, and interacting are valued as the apps progress.

Haptic User Knowledge

Having interrogated the assumptions that haptics are interfaceless and universal, students and I explored the related assumption that users' knowledge is not valuable in haptic technologies. Discourses of use surrounding haptics often emphasize that no user manual or learning curve is needed to use haptic technologies. As described, Han characterizes multitouch technology as highly intuitive and needing no manual. The highly publicized commercial for Apple's iPad also suggests this by starting with the question, via voice-over, "What Is iPad?" The commercial answers this question both verbally and visually, layering voice-over narration over multiple images of people's hands and fingers using iPads to scroll, enlarge, select, move, and manipulate texts, videos, and images. The narrator explains, "There's no right or wrong way, it's crazy powerful, it's magical. You already know how to use it. It's more books than you could read in a lifetime. It's already a revolution and it's only just begun" ("What Is iPad?"). In this definition of the iPad, users' knowledge, including any ways they may modify or improve on the iPad through use, is a foregone conclusion. As Carolyn Miller has pointed out, rhetorics of technology often make use of timing or opportunity (*kairos*) as a powerful theme. She identifies a discourse of "technological forecasting" by which "the characterization of and construction of moments in the present are crucial to the projection of the future" ("Opportunity, Opportunism, and Progress" 82). Apple mobilizes a large amount of technological forecasting in its iPad definition, collapsing current time with future predictions. In particular, the directive "You already know how to use it" signals to audiences that there is nothing new to learn or to modify. Although the statement suggesting that there is no right or wrong way to use the iPad seems to encourage improvisation and modification, it is followed by a statement that precludes innovation. If users "already" know how to use the product and if the "revolution" has "already" begun, then users are left with little incentive to modify, innovate, or even contribute to the use of the technology. Although haptics would seem to be the most hands-on of technologies—inspiring adaptation, invention, and improvisation—the rhetoric surrounding them precludes these possibilities. Users' knowledge is undervalued and dismissed in this characterization.

As Robert Johnson has shown, a user-centered approach to technology, rather than an approach focused on the designers and developers, encourages more agential and participatory relationships between humans and technologies. Crucial to this approach is recovering users' knowledge, especially their "know-how" or "doing"—a knowledge that is "not seen, not heard, and not known, a type of knowledge that has been stripped of its ability to consciously voice its purpose, power, and means by which it can make its knowledge visible" (R. Johnson 5). People with disabilities have always used technologies, including assistive technologies, in inventive and innovative ways. In their daily lives and practices involving technologies, they have often repurposed certain technologies, remaking them for different contexts of use, improvising *kairotically* based on changing situations. Assistive technology users in the public eye such as Aimee Mullins frequently draw on themes of "developing the innovation habit" in public speaking appearances (aimeemulllins.com). A visit to any current Web site for assistive technologies, such as rehabtool.com, shows people with disabilities swapping hints, tips, and tricks regarding how to make the best and most flexible use of technology. This knowledge, as Johnson suggests, is often neither seen nor heard but is often made obvious through practice and tacit knowledge.

Although discourses of technological forecasting in haptic technologies have attempted to devalue user knowledge, I led students toward examining examples of users interacting inventively with technologies. We viewed the film *Murderball,* a documentary of a quadriplegic rugby team, and I directed students to note how the players modify their wheelchairs in innovative ways and resist stereotypes of people with disabilities. Students and I discussed a particular scene in which Mark Zupan, the captain of the rugby team, visits Keith Cavill, a newly paralyzed man in a rehabilitation hospital. Cavill, who had been paralyzed in a motocross accident, pleads with hospital staff to let Zupan use his modified wheelchair to tap lightly on his own more traditional wheelchair to simulate a rugby "hit." Staffers deny this request because of liability concerns, but Zupan and Cavill clash gently anyway. The camera focuses on the tap of the wheelchairs and the expression on Cavill's face as he begins to realize that he can continue his interest in sports and athleticism as a wheelchair user. This scene is followed by another in which the rehabilitation staff present a card to Cavill upon his departure. Cavill has difficulty opening the envelope because of the limited dexterity in his fingers. The camera records emotions surfacing on staff members' faces as they debate whether to offer help to Cavill. He soon opens the card independently and appears satisfied. In

Cavill's emotional departure from the rehabilitation hospital, students noted that he looks happy, proud, and relatively self-assured, in contrast to how he seems just a few scenes later when he arrives back home and investigates the modifications to his house in progress for wheelchair accessibility. He approaches the semioutfitted bathroom skeptically, registering it as an obstacle. Students noted that in this unsure and transitional stage of his recovery, Cavill's contact and connection with the rugby team, both physical and emotional, are crucial.

I drew attention to these scenes in relation to Linton's project of "reassigning meaning" in disability studies, emphasizing how user-centered approaches to assistive technology resist medical meanings of disability. Linton, interested in how linguistic conventions construct people with disabilities as passive objects, critiques phrases such as "*wheelchair bound*" or "*confined to a wheelchair,*" writing that "disabled people are more likely to say that someone *uses* a wheelchair" because this refers to the "active nature of the user," recognizes "the positive way that wheelchairs increase mobility and activity," and emphasizes that "people get in and out of wheelchairs for different activities: driving a car, going swimming, sitting on the couch, or, occasionally, for making love" (*Claiming Disability* 27). As a class we discussed the scenes in the film in which the rugby players describe how their wheelchairs have been modified to withstand the hard-hitting and physical contact of the sport. Students also noted the agential ways in which players manipulate their chairs, knocking opposing players down and barreling down the court. I encouraged students to see assistive technologies as technologies that are used by active users. I drew special attention to the fact that the rugby players' modifications of the "typical" wheelchairs into rugby wheelchairs demonstrate that all technologies, whether assistive or not, are always ripe for adjustments, adaptations, or repurposing by users at various stages.

I challenged students to apply their valuation of users' knowledge in *Murderball* to their study of haptic technologies. A small group of students led a discussion based on modifications that people with disabilities have made to the iPad. They focused on a blog called "Life; Paralyzed" and a review of the iPad by a woman with significant mobility impairments. They noted that the user, Christina Symanski, illustrates an active approach to using the iPad, and they discussed the ways she has made it more accessible for her specific needs while offering suggestions of ways that Apple can improve accessibility. As Symanski relates:

> Overall, the functionality is pretty intuitive and I can do most things that anyone else could. There are a few multi-finger gestures that I can't do, but for the most part, I've been able to navigate quite well, using my stylus. Certain functions can be done in multiple ways, for example, you can enlarge or shrink pages by pinching and pulling at the corners or you can double tap. I have noticed that the iPad

> is very particular and is programmed to sense tiny differences in movement, pressure and the duration of your finger or stylus is [sic] on the screen. Because it is so sensitive, it does seem to work more accurately and consistently with fingers, versus the stylus. There have been many times that I've been using it and I have to try to push the same area two or three times, before I get it just right. It can be annoying at times. I think the easiest solution would be, for Apple to design a capacitive stylus with a more precise tip. All of the capacitive styluses that I've come across (including the Pogo that I'm using) have a thick tip, about the circumference of a pencil eraser. If they could somehow narrow the tip to a point, it would make it easier to be accurate. ("iPad Review")

Students noted that Symanski, like Ziegler, resists the dominant cultural logic of the universal, able user of haptics. Symanksi asserts that "certain functions can be done in multiple ways" and devises movements that work for her mobility even if these do not match those of most users. Furthermore, Symanski resists the dominant cultural logic of haptics that devalues users' intervention and expertise. Students complicated the Apple directive of "you already know how to use it" and explained that Symanski's knowledge of how to use the iPad derives from her experience living as a person with paralysis. In fact Symanski draws on this knowledge to suggest changes that Apple might make in the future to increase accessibility for more people.

Group assignments such as these prepared students for their larger individual projects. I directed students to choose a haptic technology to study, asking them to investigate its marketing and advertisement strategies and to explore its accessibility, usability, and design. Each student spent several weeks familiarizing him- or herself with a specific haptic technology and used our critical haptics history as a guide for identifying the ways in which specific cultural logics of haptics persist. Students typically found that new technologies continue to draw on the assumption that touch is an "interfaceless" medium between bodies and machines. They also noted how new and developing haptic technologies frequently assume and construct an able or universal user and devalue user knowledge and know-how, but they were also able to find examples of variously able-bodied and disabled users who counter these assumptions with their diverse range of user knowledge.

After investigating a specific cultural artifact of haptics more closely, students demonstrated increased cultural literacy regarding the study of touch and haptic technologies. Contrasting their understanding of touch before their chosen project, several reported that touch had been a sense they had "taken for granted," "not thought much about," or even "ignored." Having explored the sense of touch more fully in relation to technology, however, many students reported increased awareness of the social and cultural meanings of touch, including how haptics contributes to our desire for a "sense of immediacy" as a culture, a search for the "perfect embodied connection" between ourselves and technology, and occasionally to our "overreliance" on a "superficial" sense of intimacy and connection. Several students were critical of the advertising and marketing of some haptics that position touch as a "novelty" or "oversimplified," when in actuality touch is a sense with deep cultural meanings and resonances that can be mined for limitless approaches to technology. The experiences of people with disabilities interacting with and modifying technologies exemplified avenues for exploration for students, whether they identified as nondisabled, disabled, or temporarily able-bodied. Students also found possibilities for potential new directions for their future interactions with haptics, describing touch as a sense that makes them feel "powerful" and "in control" of their connection with technology.

Haptic *Logos,* Haptic *Mētis,* Haptic *Kairos*

Pedagogically, I consider the teaching of a course unit on haptics as a potential application of the general aims and objectives of the study of rhetorical touch. I aligned course objectives with the lessons that attending to touch as a rhetoric teach, while being mindful of the quickly changing nature of technological research and development. Students learned that the logic of touch is not as simple and clear-cut as haptic technologies suggest. Touch does not necessarily provide unmediated experience or dissolve the boundary between body and machine. In resistance to the assumption that users are able and universal, students learned that a wide range of nonnormative bodies and minds interact productively with touch and haptics.

Creators of technologies historically construct touch as the ideal interface and as the barrier to be removed between users and machines, but this contradictory logic only stifles productive engagement about the nature of the interface. The nature of the interface between bodies, technologies, and media has long occupied critical thinkers and questioners. I consider students, to some extent, to be furthering the questions that both Empedocles and Aristotle asked about touch and the interface. Rather than foreclosing questions about the blurred boundaries between self and other, mind and machine, and body and technology that haptics raise, cultural discourses should invite investigation. Students, in their individual projects on specific haptic technologies, supported with class discussions and small group projects, learned how to ask these crucial questions. The questions they ask contribute to the construction of what Stuart Selber calls the "rhetorical literacy" of interface design, in which students practice and understand persuasion, deliberation, reflection, and social action (144–66). Students' literacies reveal that multiple logics circulate in the production, design, and use of haptic technologies. These multiple logics are based on context, situation, identification, and the interests of various audiences and stakeholders.

To navigate a digital world, nondisabled and disabled students must possess flexible thinking styles, information savvy, and multisensory and multimodal composition strategies. They must be able to craft arguments in a variety of diverse media, logics, and modes that draw on all the senses. It is no coincidence that the icons used to represent a useful Web browsing experience, from the fox of Firefox to the ship of Netscape Navigator, implicitly draw on associations with models of *mētis.* Even the metallic material of the Google Chrome browser icon recalls an ancient and flexible material of *mētis.* Students must operate as *mētic* information hackers in today's digital society; they must *kairotically* know not only how but also when to use technology, depending on situation, timing, context, and other factors. These desired skills and literacies demand haptic *logoi*—pliable thinking and cognition styles. These abilities, however, do not necessarily depend on traditional notions of ability. In fact people with disabilities are often in possession of the kind of diverse, multimodal, and multimediated strategies that digital culture demands. They are frequently able to move flexibly among media, modes, technologies, and thinking styles in response to changing and diverse situations. Disabled people who draw on all their senses to negotiate rhetorical situations, facilitate rhetorical identifications, and interact with technologies resist the tradition of the perfectly able rhetor, modeling possibilities for new types of rhetors and students of rhetoric.

I consider students who are investigating haptics to be working with models of thinking and doing that are similar to the models of thought and action displayed by the tactile actions of figures in rhetorical theory such as Athena and Hephaestus. Athena and Hephaestus model a kind of haptic *mētis,* displaying an attention to touch based on difference, flexibility, and nonnormative embodiment and cognition. Their tactile interactions, at the boundaries of bodies, technologies, and machines, model strategies for productively interacting with haptics. Athena's facilitative *mētis,* operating at the side of multiple bodies, and Hephaestus's more hands-on *mētis,* functioning directly with bodies in motion, offer models for the wide range of bodies and cognitions that interact with haptic technologies. They offer models of a cunning approach to haptic technologies, demonstrating that there is no one "universal" approach to technologies, but rather that all approaches to technology, assistive or not, are like *mētis*—multiple and diverse, situated in the context of dynamic, changeable, and shifting situations.

In resistance of the assumption that users do not have any knowledge to bring to the design and use of haptic technologies, I consider students putting into practice a kind of haptic *kairos* in which users bring innovation at every stage of the technological process. Like the girl operating Empedocles's water clock or the holder of the shield being struck in Aristotle's example, students learn to be attentive to touch, to tease out its complexities, and to interact agentially with technologies. Students imagine the wide range of ways that technologies, assistive or not, can be repurposed, discovering people with disabilities actively engaging with and modifying haptic technologies for their own particular uses and situations. In this way students understand that haptic technologies are also *kairotic* technologies, ripe for revision according to context, timing, and situation. Opportunities always exist for contributing user knowledge to technology, and the experience of disability often occasions useful changes, adaptations, and modifications. Students realize that all technologies, whether specifically designed for people with disabilities or not, assist in the function of daily life, and they look for ways to engage actively with them rather than accepting them uncritically.

Conclusion

Holding On and Letting Go—Toward an Ethics of Rhetorical Touch

In 2002 laboratories at Massachusetts Institute of Technology and University College London collaborated to accomplish the first transatlantic "virtual handshake." Using PHANToM haptic (touch-based) technologies, engineers and researchers working across an ocean applied principles from telerobotics experiments and rehabilitation technologies designed for people who are blind and visually impaired to expertly simulate the minutest details of a typical handshake, calibrating the correct length, pressure, and angle (Wang et al.). The haptic technologies used to create this handshake are similar to the technologies currently in everyday use, ranging from iPads and iPods to cell phones, computers, cars, and video gaming systems. As the virtual handshake suggests, haptic technologies such as these, especially when combined with mobile communication capabilities, have the potential to become something akin to second skins for users. As David Howes notes, it "has been argued that the inhabitants of the contemporary West and other 'wired' societies have acquired new, electronic, skins, the skins of televisions and computers, which relay information about the world to us" (30). The virtual transatlantic handshake between MIT and UCL, enabled by haptic technologies, certainly signals a new approach to communication technologies ahead, as second skins such as those used via haptics enable users to become more connected to each other in an increasingly technologically visual and tactile world.

Additional futuristic haptic encounters are already in the making, many of which raise ethical questions about the uses to which touch is put. As Nina Jablonski explains, in the future touch will not be restricted by physical proximity, as haptics will enable tactile interaction from remote locations. She describes the variously abled humans, animals, and technologies that will populate the future landscape of haptic communication technologies: "think, for example, of providing a soothing touch to a group of invalid elderly individuals, offering a long-distance hug, enjoying a remote interactive grope with your latest chat-room hookup, or engaging in a bit of harmless virtual bondage. These scenarios are closer to becoming reality than you may think, as the capacity to deliver a 'cyber caress' to an animal now exists, paving the way for a 'hug suit' for humans that can deliver a soothing hug to another person via the internet" (171–72). Jablonski warns that the same technologies that might offer a soothing hug or an inviting handshake also have the potential to do severe harm. "The same technology will, of course, also open the door to more sinister applications, such as espionage, abusive interrogation, virtual molestation, and torture. The possibilities are both captivating and frightening and are likely to challenge our basic notions of self, physical presence, and personal responsibility" (172). This challenge necessarily involves asking questions about the role of touch not only in daily life but also in our histories, theories, practices, and ethics.

A response to the virtual transatlantic handshake and the future it signals might consider the ancient philosopher Zeno's analogies of the closed fist and the open hand to symbolize logic and rhetoric. As Edward P. J. Corbett explains, the "closed fist symbolized the tight, spare, compressed discourse of the philosopher," while the open hand stood for the "relaxed, expansive, ingratiating discourse of the orator" (288). In a work written at the end of the tumultuous 1960s, Corbett compares then-current examples of rhetoric of the open hand to "reasoned, sustained, conciliatory discussion of the issues" and rhetoric of the closed fist to "non-verbal," "provocative" means of communication, otherwise known as a "body rhetoric" or "muscular rhetoric" (288, 291). Corbett's suggestion for bringing these two modes of rhetoric together is to think of them as of a piece with one another, occupying the same body. He explains, "The open hand and the closed fist have the same basic skeletal structure. If rhetoric is, as Aristotle defined it, 'a discovery of *all* the available means of persuasion,' let us be prepared to open and close that hand as the occasion demands" (296).

The occasions demanded by the development of haptic technologies in the twenty-first century necessitate a flexible rhetoric of both the open hand and the closed fist. This flexibility is especially crucial for garnering a response to ethical questions and entanglements such as those that Jablonski notes when cataloging the various kinds of haptic technologies developing on the not-so-distant horizon. Although it may be tempting to read Corbett's assumption of the closed fist and the open hand as having the "same basic skeletal structure" as normative or even as exclusively able-bodied, it is more generative to think of the open hand and the closed fist in both symbolic and literal ways. Symbolically the flexibility of the same hand using both the open hand and the closed fist signals a willingness to engage with questions of ethics and limits in the area of haptics development. More literally both the open hand and the closed fist, no matter what their shape, are a reminder that haptic technologies have real, material effects on the bodies—disabled and nondisabled—that interact with them. In both symbolic and material registers, the ethical questions that emerging haptic technologies occasion can be informed by a study of the sense of touch as a rhetoric. In some cases the ethical questions of haptics may invite an approach of the open hand; in others, especially when issues of power and domination are involved, the closed-fist approach may be necessary.

In this book I have explored touch as a rhetorical strategy that possesses a history, theory, and practice. Telling stories about touch—histories, rhetorics, current practices, personal and shared experiences—is the beginning of an ethical response to the challenging questions that touch and haptics raise concerning basic notions of self, other, mutuality, and responsibility. The next step in an ethical response requires asking crucial questions. If touch is truly a *dynamis,* a potential for grasping meaning in new ways, then what must we hold on to and let go of to fully realize touch as a rhetorical and ethical art? The *dynamis* of touch encourages us to let go of some of the negative associations between touch and rhetoric in the tradition while remaining aware of the potential for misuse. Touch also invites us to more comprehensively let go of the idealized image of an independent, nondisabled, singular rhetor and to embrace the possibilities of new configurations among communicators, speaking and nonspeaking, and multiple audiences across ranges of ability and disability.

These new configurations necessitate a holding on to the responsibilities of a rhetoric of touch and an examination into the province of its ethics. At the core of ethical engagements enacted by touch are feelings of mutuality, respect, and esteem that circulate not in spite of but because of differences in how one inhabits the world. Those who enact ethical engagements through touch do so by holding on to the radical complexity and uncertainty that touch occasions while letting go of the limits often imposed by assumptions based on disability, language use, or even personhood. A rhetoric of touch rooted in these engagements begins to formulate a foundation for approaching the ethical questions raised by touch in everyday life and in a future of haptic technologies. This ethical foundation involves considering the positive and negative aspects of touch, its complexity and potential for connection across ranges of abilities, and its aptitude for working at the edges of language, particularly in relation to our environment, companion species, and the widest ranges of our relations.

The Tangles of Touch

The experiences of people with disabilities are clearly invaluable for understanding the potential of touch as an ethical rhetoric. Disabled rhetors who use touch rhetorically frequently do so in an ethical register. People with a wide range of sensory, cognitive, psychological, and physical disabilities negotiate touch to establish themselves as rhetors, to connect emotionally to other people, and to craft persuasive messages. Touch, they have demonstrated, can be a vehicle for appeals based on *logos, ethos,* and *pathos.* In fact touch more often than not results in the blurring of boundaries between these seemingly self-contained elements of rhetoric, and it encourages readings of rhetorical situations, appeals, and identifications that seep into each.

The efforts of disabled people who use touch rhetorically indicate even more seepages than I have examined, demonstrating the wide range of ethical applications to which a rhetoric of touch can pertain. I have focused on the *logos* of people with schizophrenia and depression, but rhetors with psychological disabilities, in forming appeals based on a sense of felt *logos,* also create ethical characters full of goodwill and credibility, despite assumptions that they are unreasonable, unscrupulous, or unstable. Although I have focused on the *ethos* of people with autism, autistic rhetors also use touch to craft emotional appeals of *pathos* through their characters, in resistance to stereotypes that they are cold, emotionless, or purely literal-minded and logical. I have focused on the emotional appeals of people with physical disabilities, but these appeals emerge from other contexts of resistance in which people with disabilities counter limiting messages about disability, crafting appeals of *ethos* and *logos* in difficult situations. Disabled rhetors who use touch to communicate or to augment communication make ethical interventions most broadly by resisting stereotypes about disability. They often reset the terms of the discussion about disability by presenting opportunities to reevaluate assumptions and to generate new knowledge about disability.

A rhetoric of touch necessarily involves ethical considerations because issues of vulnerability, trust, contingency, and accountability are all intertwined in touch. We touch and are touched; this dual relationship generates meaning, engagement, and identification. On the one hand, touch is generative and productive, establishing context, marking relationships, inviting identifications, and often communicating meaning and feeling at the edges and limits of language. On the other hand, however, touch can be anxiety-provoking and dangerous. In his exploration of social touching, Stephen Thayer expresses caution about touch: "Touch represents a confirmation of our boundaries and separateness while permitting a union or connection with others that transcends physical limits. For this reason, of all the communication channels, touch is the most carefully guarded and monitored, the most infrequently used, yet the most powerful and immediate" (298). As Thayer suggests, there is an impulse to guard and monitor touch as a communication medium. We are often suspicious of touch, aware of its power, but hold it in abeyance. As much as touch supports connection, enables identification, and circulates emotion, it also has the potential to do harm, to injure, and to divide. I have chosen to focus on the most generative and productive of touches, but undeniable is the darker side of touch—abuse, violence, pain, and force. Examination of this other side of touch is particularly relevant in the context of disability, as people with disabilities are far more likely than those without disabilities to be abused in a variety of situations. Although I have not taken up this examination in this investigation, power, domination, and control are as much parts of touch as are trust, identification, and fellow-feeling. An ethics of touch must be responsive to these tangles of touch and especially mindful of the impulses both to guard touch and to realize its power.

Even in the contexts of the most productive identifications made through touch, there are ethical considerations that must be taken into account. I have focused on the possibilities afforded by identification and partial identification among people with disabilities and their various audiences, but there are also risks in this process. In rhetorical studies, theorists such as Krista Ratcliffe have noted the limitations of identification as a metaphor of common ground that connects people. Ratcliffe suggests that current theories of identification often do not "adequately address the coercive force of common ground" (47). As Cynthia Lewiecki-Wilson explains, providing an example, "When I assert that I identify with you, I am in a way stating I am you, and so potentially I enact a violence, an erasing of our very real differences" ("Ableist Rhetorics, Nevertheless" 81). In the context of disability, this is particularly problematic. Identifications made among disabled and nondisabled individuals, although productive in some situations and necessarily partial and contingent in other contexts, must always be sensitive to the risks of erasure in identification. There are significant material differences between living as a disabled person and living as a nondisabled person, and there are also important experiential differences in living as a person with a physical disability and living as a person with a cognitive or psychological disability; these differences are important to value because the experiences of all disabled people are crucial repositories of meaning-making. Transcending or overcoming the real differences between people with disabilities and nondisabled people or among people with different kinds of disabilities should not be the objective of identification, however partial or contingent. Instead the objective of identification should be to create potential spaces for flexible connections between people with disabilities and a wide range of audiences and collaborators—people with different disabilities, people with similar disabilities, those temporarily able-bodied, and nondisabled people.

When used in an ethical manner, rhetorical touch acknowledges the risks of identification but also capitalizes on its potentials. Ratcliffe suggests the figure of metonymy to augment identification, noting that metonymy "foregrounds resemblances based on juxtaposed associations, thus foregrounding both commonalities *and* differences" (68). Since metonymy operates via association and resemblance rather than direct substitution or erasure, it can accommodate the paradoxes inherent in different identities in identification. As Lewiecki-Wilson explains, metonymy potentially contains "the paradox that identity includes non-identity within it: I may be like you in some ways, share some associations, but I am not like you in other ways" ("Ableist Rhetorics, Nevertheless" 81). Understanding touch as a type of metonymy can extend this potential further. Touch is a temporary, contingent, and flexible location for association that can lead to potential identification among individuals.

The mental disability theorist Margaret Price models an expression of touch as a metonymic type of identification in her telling of a story about her relation to the disability community that became particularly clear during an experience at an academic conference:

> On a December day in 2008, I arrived in a fluorescent-lit hotel room in San Francisco to listen to a panel of scholars talk about disability. I had recently made a long airplane journey and felt off-balance, frightened, and confused. I sat beside disability activist and writer Neil Marcus, and when he saw my face, he opened his arms and offered me a long, hard-muscled hug. That hug, with arms set at awkward angles so we could fit within his wheelchair, with chin digging into scalp and warm skin meeting skin—that, to me, is disability community. Neil may or may not know what it is to wake with night terrors at age forty, I may or may not know what it feels like to struggle to form words, but the reaching across spaces is what defines disability for me. We write, we question and disagree, we are disabled. (*Mad at School* 20)

For Price and Marcus, touch operated as a metonymic identification in a specific place and time. Through a hug, they exchanged connection that included both identification and difference. For Price, this hug also functioned rhetorically, as a "reaching across spaces" that defines disability and provides impetus for writing, questioning, and experience. This hug models a particularly effective rhetorical and ethical approach to touch, functioning as a way of building connection on individual and community levels without erasing material difference. Touch deindividualizes disability while simultaneously enabling recognition of differences among individuals.

Touch beyond Boundaries

Ethical interventions by people with disabilities who use touch as a rhetorical strategy model approaches to identification across traditional barriers of ability and disability without diminishing the material differences between these categories. Rhetorical touch used in an ethical manner does not overcome differences between people with disabilities and nondisabled people or transcend differences between people with different disabilities. It is tempting to understand touch as a sense that equalizes; yet it is crucial to be mindful of its differences. Unlike the other senses, touch is almost never completely disabled. Except in the case of a rare neurological condition, touch "is present within every single interaction with objects" in the world and also occurs within a considerable amount of interactions with other people (Paterson 2). People with a wide range of disabilities, nondisabled people, and temporarily able-bodied people can, in different ways, typically all draw on touch as a way to communicate, interact with the world, and forge identifications across traditional barriers. Ralph James Savarese realizes this potential while using a form of tactile therapy with his son, who has autism: "All of us, I would say, whether disabled or not, are dying to be touched; [although] it's hard to quantify the benefits of such touching" (269). These benefits are diverse, multiple, and context-dependent.

Rhetorical touch, operating in an ethical register, reveals opportunities for emotional and embodied identification not in spite of but because of the differences attributed to various forms of ability and disability. In the emerging interdisciplinary field of critical affect studies, there is potential for exploring how touch, in physical, emotional, and discursive registers, transmits feelings among bodies of varying abilities. Eve Kosofsky Sedgwick explores a potential of touch in relation to affect in her text *Touching Feeling,* a title that "records the intuition that a particular intimacy seems to subsist between textures and emotions" (17). This particular intimacy is represented by the "double meaning" that is already present in "the single word 'touching'" and its counterpart "feeling" (17). Sedgwick experienced the "touching feeling" that affect creates when viewing a photograph of the outsider textile artist Judith Scott hugging a piece of her own work, a tactile conelike assemblage of yarn, cord, ribbon, and other fibers. Scott, who has Down syndrome and is deaf, does not use language in traditional ways and spent much of her life warehoused in an asylum before becoming known for her art. Sedgwick muses about the connection she feels toward the photograph: "I don't suppose it's necessarily innocuous when a fully fluent, well-rewarded language user, who had never lacked any educational opportunity, fastens with such a strong sense of identification on a photograph, and oeuvre, and a narrative like these of Judith Scott's. Yet oddly, I think my identification with Scott as less the subject of some kind of privation than as the holder of an obscure treasure, or as a person receptively held by it" (23–24). Sedgwick notes the gulf of identification that would seem to exist between herself, a particularly able "language user," and Scott, a woman who is considered languageless. Yet Sedgwick also expresses fascination by how strongly she "fastens" on with such a powerful sense of identification with the photograph. Treasuring this fascination, she writes that she seeks to hold on to this feeling of identification in disidentification as much as to explore it and to value its complexity and inexplicability.

Rhetorically, Sedgwick identifies with Scott because of what the photograph communicates to her and about her—Scott's "artist's ability to continue asking new, troubling questions of her materials that will be difficult and satisfying for them to answer" (24). The photograph also conveys to Sedgwick "an affective and aesthetic fullness that can attach even to experiences of cognitive frustration" (24). Toward the end of her life, Sedgwick identified with Scott in the hopes that she could relate this cognitive frustration to the experience of the "senile sublime," a notion articulated by Barbara Herrnstein Smith and to which Sedgwick was also attracted. Writing at the end of her life, Sedgwick admitted that articulating herself in *Touching Feeling* had meant dealing repeatedly with cognitive frustration completely unlike what she imagined Scott may also have felt. She explains, "In writing this book I've continually felt pressed against the limits of my stupidity, even as I've felt the promising closeness of transmissible gifts" (24). Sedgwick asks difficult questions of the materials of language, communication, and affect not unlike how Scott asks difficult questions of the materials of her sculpture. Sedgwick recognized and felt with Scott the effects and affects of this stream of difficult questions. Perhaps most tellingly, she writes that she "identif[ies] with the very expressive sadness and fatigue in [the] photograph" (24). Sedgwick was able to connect with Scott not through words but through emotion, touch, and fellow-feeling. Her identification was not based on erasure of Scott's experience and difference but rather rooted in shared associations and metonymic resemblances.

Touch is a sense that productively blends with other senses, including the sense of sight, creating new opportunities for connection in identification, justice, and ethical obligation. Rhetorically, Sedgwick identified with Scott through an aesthetic response that was both tactile and visual. Sedgwick also responded aesthetically to Scott's art; aesthetics, as the disability theorist Tobin Siebers relates, "tracks the sensations that some bodies feel in the presence of other bodies" (*Disability Aesthetics* 1). To a certain extent Sedgwick "fastened" on to the visual of Scott in an act of "beholding," a visual experience that also involves physicality. Rosemarie Garland-Thomson, in her exploration of the act of staring, uses the psychologist D. W. Winnicott's concept of the holding function, which is experienced in childhood by the mother's holding gaze as well as by the literal holding of the infant close to the body. Assuming that "to be held in the visual regard of another enables humans to flourish and forge a sturdy sense of self" and that "being seen by another person is key to our psychological well-being," Garland-Thomson writes that staring "might be understood as a potential act of be-holding" (*Staring* 194). In this act the "work of a beholding encounter would be to create a sense of beholdenness, of human obligation, that inheres in the productive discomfort [that] mutual visual presence can generate" and to initiate the "opportunity for generating mutual new knowledge and potential social justice" (194). Beholding can be a tactile as well as a visual act, with the potential to generate productive and mutual knowledge. Touch, like sight, countenances a response. In touch we also sense each other, register self-worth and mutual esteem, and exchange information. Sedgwick and Scott, for example, did not share a language in the traditional sense, but they shared a system of exchange and response through aesthetic, affective, and rhetorical registers, communicating connection and opportunities for mutual esteem. Sedgwick's identification with Scott did not erase the differences between them but did create a space for engagement in an ethical register in which Sedgwick was beholden to the questions that Scott's work raised for her and valued the possibility for mutual knowledge between them.

For rhetors who do share language in the traditional sense, identifications afforded by touch sometimes result in shared communication or writing styles. Janet Price and Margrit Shildrick, for example, have forged a writing partnership based on touch. They imagine that one alternative to binaries that privilege either the material or discursive experiences of embodiment is "to write together where the question of unique authorship—and thus of authority to speak—is suspended" (63). Price and Shildrick, by writing together as one disabled woman and one nondisabled woman, reject both the "suggestion that disability is not an issue for nondisabled people" and "that there is some privileged standpoint from which disabled people alone can speak" (64). This decision to write together grew out of the embodied partnership that Price and Shildrick cultivate in their friendship. They explain: "What we want to emphasize is that the willingness to give up ownership of the text parallels the willingness to give up ownership of 'my' body, which in each case opens up new social and ethical possibilities" (65). Identifying that in most typical single-authored texts about embodiment and disability "what seems to be missing is an ethical dimension that moves beyond a simple recognition of vulnerability in the self," they offer instead a "recognition of difference and vulnerability in others" (66). This recognition arises from lives lived in contact and relation. Writing temporarily from her singular perspective, Price explains, "Security from friends grows from the prior knowledge of what the embodied experiences of self and other have in common: hands held, coffees drunk together, dances and hugs, the afternoons in the sauna or spa pool, the mutual touches of friendship, of comfort, of joy, of sadness translated into the moves and actions needed to lift me, to help me transfer, to climbing in with me to help me bathe. These friends do not experience my body as it is now for me, but our mutual histories offer an ease through these moments" (71).

Shildrick also contributes her own perspective, writing, "As a result of Janet's overtly fluctuating sense of touch, in which her personal body map was clearly undergoing unpredictable changes, I (Margrit) was also unsettled in the context of tactile interaction. Lacking a clear sense of Janet's own corporeal boundaries, I found it difficult to know at any given moment whether a greeting hug was still experienced as a sign of affection, given and returned, an unfelt and therefore meaningless gesture or even a more or less painful assault. Although at that time

I was familiar and confident with the procedures of helping Janet wash, eat or dress, for example, those could be done with a certain distancing efficiency which deferred a recognition of the mutuality of touch" (72). Neither Price nor Shildrick seeks to erase the differences between them, felt especially in their daily physical interactions. The confusion that both Price and Shildrick experience in their tactile encounters in the wake of disability is precisely what they hold on to as they continue to inhabit each other's worlds and to cultivate a friendship based on mutuality and respect. Writing again from a joint perspective, they assert that "what was uncovered during that acute period of Janet's illness was that, through the mutuality and reversibility of touch, we are in a continual process of mutual reconstitution of our embodied selves" (72). The uncertainty and "undecidability" that Price and Shildrick experience through touch mean that "the distinctions between self and other are never firm or final" (73). For Price and Shildrick, touch is an ethics based on mutuality, respect, and responsibility. Practicing this ethics of mutuality means writing together, using language that grows from the tangible materiality of their lives. Neither Price nor Shildrick uses language based on touch to erase the material differences in their lives; rather they use it to explore partial identifications, contingencies, and disidentifications. As with Sedgwick and Scott, Price and Shildrick use touch to create a space in which identification and disidentification intertwine.

The Potentials of Touch

Connections among Scott and Sedgwick and Price and Shildrick circulate primarily through the appreciation of art and writing; yet for others, identifications can also take place beyond linguistic, visual, or discursive registers. Beyond language, touch can also create ethical connections not only among people but also among people, animals, technologies, and the various nonhuman entities of their environments. Helen Keller, contrasting the ways in which disabled and nondisabled people use touch, suggests that touch can be relevant, regardless of disability, in a wide range of ecologies: "Touch brings the blind many sweet certainties which our more fortunate fellows miss, because their sense of touch is uncultivated. When they look at things, they put their hands in their pockets. No doubt that is one reason why their knowledge is often so vague, inaccurate, and useless. . . . There is nothing, however, misty or uncertain about what we can touch. Through the sense of touch I know the faces of friends, the illimitable variety of straight and curved lines, all surfaces, the exuberance of the soil, the delicate shapes of flowers, the noble forms of trees and the range of mighty winds" (*The World I Live In* 30–31). Keller invites people without disabilities to take their hands out of their pockets and to experience the layers of meaning and wide ranges of knowledge that accompany touch. Touch is valuable, in Keller's estimation, because it translates across abilities and provides information that is productive. Keller, for example, draws attention to the ways in which touch enabled her to commune with the natural world. Touch, for her, was a potential way of knowing the flowers, soil, trees, and wind. At the limits of language, touch reveals possibilities for translating meaning beyond words.

As such, touch and its attendant meanings can function as a gateway for exploring one's relationships with elements of the natural and material worlds with which one interacts, including technologies. For some people, one's relationship to assistive technologies is structured by touch and other embodied experiences. As one woman relates: "When I use crutches I feel connected to them as if they are one of my limbs. When I set them aside and get on the wheelchair I disconnect from the crutches and take on the wheelchair as part of me. I don't like anyone to move my crutches without my permission, or to lean on the arm of the wheelchair or rest their hand on the back of it unless it is someone I feel comfortable touching me" (Olkin 23). In this instance the rules that govern who touches one's body are the same rules that govern who touches one's assistive technologies. Touch is the organizing principle for how one interacts with her technologies and for how one expects others to interact with her. Touch articulates a meaning beyond words that structures relationships among people, technologies, and spaces.

Exploring the meaning beyond words is one potential way to respond to recent calls by scholars in diverse fields to examine avenues of connection in regard to relationships with the natural world. Malea Powell uses the phrase "all our relations" to encapsulate "an entire philosophy of humans in relation to other living things—plants, animals, rocks, earth—that emphasizes the intricately connected web of relationships that sustains our mutual ability to live out our shared existence on the earth together" (Powell "All Our Relations"). Similarly, Siebers's study of aesthetics, the circulation of feeling among bodies, assumes that "bodies" include "human bodies, paintings, sculpture, buildings, the entire range of human artifacts as well as animals and objects in the natural world" (*Disability Aesthetics* 25). The philosopher Martha Nussbaum has attended to questions of ethics and social justice at the edges of language, constructing a theory of human justice to accommodate disability, nationality, and species. For both humans and animals, Nussbaum lists the "Senses, Imagination, and Thought" as values central to the capabilities approach, partially derived from Amartya Sen, which articulates a measure of opportunity based on aptitude rather than reality (75, 396). For humans, "being able to use the senses, to imagine, think, and reason" are key parts of a theory of social justice, including "being able to have pleasurable experiences and to avoid nonbeneficial pain" (76). For nonhumans, this translates into "ensuring [animals'] access to sources of pleasure, such as free movement in an environment that is such as to please their senses" (396). Regardless of ability to use language, the senses are crucial elements in the realization of a society and a natural world based on justice.

Exploring the role of the sense of touch in the construction of relationships among people, technologies, the environment, and animals can contribute to the project of making a more ethical society. Donna Haraway uses touch to ask crucial questions about the potential relationships among species and to begin to imagine a society based on new terms. She begins *When Species Meet* by asking, "Whom and what do I touch when I touch my dog?" (3). Touch contributes to what Haraway calls "becoming with" other species and "becoming worldly" with one's entire environment. This process involves a tangible kind of ethics. Haraway relates, "My premise is that touch ramifies and shapes accountability. Accountability, caring for, being affected, and entering into responsibility are not ethical abstractions; these mundane, prosaic things are the result of having truck with each other" (36). For Haraway, asking crucial questions about relating to different species and becoming accountable with the larger world mean addressing connection via touch. In the case of dogs and other companion species, this means realizing that animals and humans, in their coevolution together, are tangled up in the messiest of realities, the understanding of which is at stake in the specific question of valuing who and what inhabit the world (19).

Like Sedgwick and Price and Shildrick, Haraway mobilizes touch to theorize an ethics based on respect and a mutual sense of accountability. Investigating the material and semiotic foundations of *species,* Haraway writes that "the Latin *specere* is at the root of things here, with its tones of 'to look' and 'to behold'" (17). Accordingly, "Looking back in this way takes us to seeing again, to *respectere,* to the act of respect. To hold in regard, to respond, to look back reciprocally, to notice, to pay attention, to have courteous regard for, to esteem: all of that is tied to polite greeting, to constituting the polis, where and when species meet" (19). Similar to Garland-Thomson's sense of beholding, Haraway's *specere* is an active sense of visual and tactile engagement and mutuality. Like Sedgwick's "fastening" and Price and Shildrick's writing together, Haraway's *specere* is an ethical engagement based on mutuality and fellow-feeling.

Two end-of-life scenes characterize the ethical entanglements based on a tactile sense of *specere* that Haraway seeks to describe. When considering her father at the end of his life, Haraway uses a cognate of *specere, respecere,* as an "ethical regard" for examining her own and her father's lives (164). Haraway describes herself and her brothers holding their father while he died, "alert and present, in our hands" (162). She brings a sense of regard and esteem to this scene by exploring how she learned "to hold in regard, to hold in seeing, [and] to be touched by another's regard" while processing his life and death (164). Haraway enacts an active sense of "holding" regard in relation to telling the story of her father's work as a sportswriter and life, a life she reviews in relation to his childhood bout with tuberculosis and the attendant prosthetics with which he lived—casts, wheelchairs, and crutches—as "his partners in the game of living well" (166). Most productively, she tells of the regard she received in return when her father accepted and respected her own companions in the game of living well—her canine partners in dog agility training, understanding it as a highly skilled and communicative sport.

Similarly, in telling a story of her mother-in-law's experience of dementia at the end of her life, Haraway brings a sense of tactile and ethical regard that stretches across language and species. When Katherine, Haraway's mother-in-law, experienced a particularly upsetting delusion, she accused her son of calling her a liar and then immediately regretted it after he left, feeling almost hysterical knowing that "she had said something terrible to Rusten, but not knowing what it was" (203). Haraway writes, "It took a long time to comfort her, holding her and rocking and telling her she did not say anything awful" (203). Looking back on the incident, she writes, "The most interesting thing, though, was not what she and I were doing, but what she and the dogs were doing the whole time she was crying and desperate for comfort and relief from feelings of guilt, shame, and bewilderment. She was on the couch, and I was kneeling below her, my hands on her knees and hugging her periodically. Cayenne slipped her body between us (she would NOT be denied). . . . She would not budge until K was calm. Roland, meanwhile, had his head inserted between me and K's lap, putting his head on her knees along with my hands and pressing firmly against her body with all his weight. He also would not budge until K was calm. K's hands the whole time were kneading the dogs' bodies, first one, then the other. She did not know what she was doing consciously, but the touch comfort among K, R, and C was stunning" (203–4).

When language began to fail Haraway's mother-in-law, the family members—both human and nonhuman—turned to touch for expression and connection. At its most generative potential, touch may lead to explorations beyond language and even beyond species. In ethical terms, touch invites us to consider different ways of communicating through language and expression even beyond the typical limits of language. This invitation means asking important questions about who and what are typically understood and valued as language users.

Asking and exploring these questions are crucial in the formulation of responses to new and developing communication technologies, including existing haptic technologies as well as the futuristic technologies described by Jablonski. The lived experiences we have with other people and other beings and entities, including nonhumans, animals, prosthetics, technologies, and even life forms on the frontiers of biomedical invention, contribute to exploring these questions as well. As Susan Squier explains, the liminal lives that exist in an "in-between or marginal zone" of life "demand our attention" in exploring the ethical questions they occasion (*Liminal Lives* 4–5). "Along with our increasing encounters with liminal lives—from adoptive embryos to transplant recipients to elderly Alzheimer's patients hoping for stem cell therapy—comes an increasingly urgent desire to understand what these new beings mean to us, socially, politically, and ethically" (8). At the root of these questions is the study of relationships. As Haraway describes, exploring the relationship between technologies and species, "Cyborgs and companion species each bring together the human and non-human, the organic and technological, carbon and silicon, freedom and structure, history and myth, the rich and the poor, the state and the subject, diversity and depletion, modernity and postmodernity, and nature and culture in unexpected ways" (*The Companion Species Manifesto* 4). Haraway "asks which of two cobbled together figures—cyborgs and companion species—might more fruitfully inform livable politics and ontologies in current life worlds," especially in the effort of guiding us "through the thickets of technobiopolitics in the Third Millennium" (4, 10). The answer to this question is not singular; it is at the very least a question of relationship. Touch is at its most basic level a relationship, and its study is crucial for examining the questions and complications of the present, the past, and the future. Only the richest of ecologies, including human, technological, and animal entities, are necessary for formulating a response to questions of what is at stake for the future.

Looking back is a way of formulating a response to the pressing questions that matter for the future. In *Nicomachean Ethics,* Aristotle entertains the question that nonhuman animals may possess human qualities, including the possibility that "some people go so far as to say that certain species of animals have prudence" or a sense of *phrónēsis* that directs their actions so as to produce "deliberation aimed at a good result" (1141a; Detienne and Vernant 316–17). These possibilities have far-reaching implications, "call[ing] into question the radical separation between man and beasts, between reasonable beings and the rest, living creatures without *lógos,* the *áloga zōia*" (Detienne and Vernant 317). This act of calling into question the most easily accepted boundaries—those between human and animal, rational and nonrational, and living and nonliving—requires a rethinking of *logos,* a reimagining of who and what count as "reasonable" or "ethical" in the widely varied and rich ecologies of the world.

The stories we tell about touch matter to these rich ecologies and model possibilities for ethical responses. Although rhetorical history, theory, and practice have focused on the most able of human language users, there are models for valuing a richer range of who and what count in the world. Empedocles's rhetoric grows out of a worldview in which humans, nonhumans, and their environments, including nonliving matter and words, all interanimate. Mythologies of figures such as Athena and Hephaestus also embody alternative approaches to rhetoric based on connection, tactility, and difference. Athena and Hephaestus interact physically with their material environments and nonhumans, including animals and rudimentary technologies. The god Kairos, although he possesses an able body, draws attention to how proximity and physical contact affect rhetoric. The stories of figures such as these are important ways of looking back, respecting, and beholding an alternative approach to rhetoric. Looking back to tell these stories about touch is crucial preparation for looking forward to value the wide range of nonnormative bodies and minds that use rhetoric and for developing ethical responses to new and emerging technologies.

Notes

Introduction

1. I use both the phrases "disabled people" and "people with disabilities" in this book. Both phrases are currently in circulation in the disability community. Person-first language such as "people with disabilities" seeks to value the person before the disability, whereas phraseology such as "disabled people" seeks to integrate disability into identity, showing that disability cannot be separated from identity, as Mary Duffy explains in *Vital Signs* or as some autistic rights activists such as Jim Sinclair and Ari Ne'eman prefer in the context of autism. Because both of these arguments are compelling, I use both phrases throughout the book.

2. Davis describes the norm as a "configuration that arises in a particular historical moment. It is part of a notion of progress, of industrialization, and of ideological consolidation of the power of the bourgeoisie" (*Enforcing Normalcy* 49). Garland-Thomson uses the term "normate," writing, "This neologism names the veiled subject position of cultural self, the figure outlined by the array of deviant others whose marked bodies shore up the normate's boundaries" (*Extraordinary Bodies* 8). Linton uses the term "nondisabled" to "center" disability and to put the usual dominant perspective in "the peripheral position in order to look at the world from the inside out" (*Claiming Disability* 13). Although they use different terms, each theorist forwards the argument that ability and disability are parts of the same overarching system. Relatedly, Linton offers a persuasive rationale for maintaining the use of the term "disability" in disability studies while defining it as a sociopolitical category:

> While retaining the term *disability*, despite its medical origins, a premise of most of the literature in disability studies is that *disability* is best understood as a marker of identity. As such, it has been used to build a coalition of people with significant impairments, people with behavioral or anatomical characteristics marked as deviant, and people who have or are suspected of having conditions, such as AIDS or emotional illness, that make them targets of discrimination. . . . When disability is redefined as a social/political category, people with a variety of conditions are identified as *people with disabilities* or *disabled people*, a group bound by common social and political experience. (12)

3. Studies of touch exist in relation to child development (Montagu; Field), politics (Manning), film theory (Marks), natural history (Jablonski), cultural history (Classen), communication (Jones and Yarbrough), philosophy (Paterson; Ross), religion/philosophy (MacKendrick), and feminism (S. Sullivan), but in rhetorical studies no comprehensive study of the sense exists, although Perl has explored a "felt sense" of writing and Kuriyama has suggested connections between ancient Greek approaches to medicine and rhetoric through the study of the pulse. In disability studies, theorists such as Price and Shildrick and Petra

Kuppers and Neil Marcus have addressed touch productively but have focused on poetic, ethical, and intimate perspectives rather than rhetorical features.

4. As Frederick Sachs writes, touch is the "first sense to ignite" and "the last to burn out: long after our eyes betray us, our hands remain faithful to the world" (qtd. in Ackerman 71). Even in the case of paralysis of one or more parts of the body, it is rare that all feeling or sense of touch is lost throughout the entire body, and there are many instances of people learning to register additional or alternative sensations in other parts of the body (Shakespeare, Gillespie-Sells, and Davies).

5. The British social model is often credited with marking the difference between disability as "impairment" and as a disabling social condition. For example, impairment can be understood as a form of paralysis that may necessitate the need for a wheelchair, but disability is the social conditions that make this impairment an issue, including lack of accessible environments, buildings without elevators, and other barriers such as negative attitudes toward disability in society and culture. The social model in general is typically viewed in resistance to the medical model, which positions disability as an individual flaw or defect to be fixed, cured, or rehabilitated. The social model has undergone numerous revisions, particularly regarding the importance of the body and embodiment in theorizations of disability. As Carol Thomas describes, the body has been "lost and found" in disability studies in myriad ways. For more information on the body in social and medical models, see Shakespeare, "The Social Model of Disability." People with disabilities often take a social or cultural approach to disability, resisting a medical perspective, and explore their disabilities and their experiences in the world as the intersections of a wide range of complex social, cultural, political, and environmental factors. See Snyder and Mitchell, Linton, and Davis, among others, for social and cultural approaches.

6. Charlton, in *Nothing about Us without Us*, recalls hearing this slogan in South Africa in 1993 from two separate leaders of Disabled People of South Africa, Michael Masutha and William Rowland, and writes that "the slogan's power derives from its location of the source of many types of (disability) oppression and its simultaneous opposition to such oppression in the context of control and voice" (3).

7. Empedocles suggests this when describing to his listeners how to know and understand God. He explains that they must use alternatives to typical methods: "It is not possible for us to set God before our eyes, or to lay hold of him with our hands, which is the broadest way of persuasion that leads into the heart of man" (Diels and Kranz 31.B.133; Burnet 225). Even though he suggests that touch is not ideal for divine knowledge, he still maintains that knowledge via the hands is one of the most useful ways of understanding in general.

8. As George Kennedy, drawing from Hesiod and Herodotus, explains, the Andrians were told that the Athenians approached them with two goddesses, Peitho and Anangke, a pairing that represented "speech to persuade them to act in their own best interests and the use of force to constrain them if necessary" (*A New History of Classical Rhetoric* 13). Previously, Peitho had been known as the daughter of Aphrodite and was associated with sex. Peitho was first explicitly identified in Plato's *Gorgias*, in which the definition of rhetoric accepted by the sophist is "*peithous dēmiourgos*," or "the worker of peitho" (13).

9. There is some debate regarding how well Quintilian's analogy endures for art lovers, but as Rollins observes, "nothing could have been less suitable for sculpture than the stone used for millstones; and Quintilian might suppose that it would have been impossible, even

for a Praxiteles, to have produced even a tolerable statue from it" (qtd. in Bizzell and Herzberg 396). Quintilian also compares speaking and art in a similar context of disability in his description of Apelles's decision to paint Antigonus "with only one side of his face toward the spectator, so that its disfigurement from the loss of an eye might be concealed. Are not some things, in like manner, to be concealed in speaking. . . ?" (384).

10. Studies of the visual in rhetoric and writing are numerous. Noteworthy examples include Bartholomae and Petrosky, Fleckenstein, George, Hilligoss, Kress, McComiskey, Odell and McGrane, Prelli, Trimbur, Tufte, and Wysocki. New Media studies in writing and composition often focus on sight and sound, for example: *Writing New Media, Remediation, Multiliteracies,* and *Multimodal Composition.* Studies of sound in writing and composition include Cynthia Selfe's "Aurality and Multimodal Composing," Krista Ratcliffe's *Rhetorical Listening,* and Cheryl Glenn's *Rhetoric of Silence,* which explore sound and its absence.

11. Valuable studies at the intersection of rhetoric and composition and disability studies often focus on a specific kind of disability—deafness, physical disability, blindness, neuromuscular disability, or mental disability, among others. Identification between bodies, especially of different disabilities or with nondisabled people, has yet to be explored comprehensively. The visually marked disabled body has been particularly well explored in both rhetorical studies and disability studies. Approaches to disability in rhetorical studies featuring visibility include those by Brueggemann, White, Dunn, Heifferon, and Cheu; Barton; Barber-Fendly and Hamel; Mossman; and James Wilson, among others. In disability studies, see Garland-Thomson, Davis, and Siebers, among others. Recently scholars such as Dolmage, Price, and Vidali have explored "invisible" disabilities in addition to visible disabilities.

12. In rhetoric and composition, various constructions of the body have been studied—the teacher's body, the student's body, the gendered body, the sexed body, the classed body, among others. See Paterson and Corning and Selzer and Crowley for valuable examples. I seek to build on these explorations of the body, shifting attention from the body to an explicit examination of bodies in physical contact.

13. As I explore more in-depth in chapters 1 and 3, the recent debates surrounding the authenticity of facilitated communication (FC) show that rhetorical productions made via touch are still doubted and dismissed. People with disabilities who form messages on computing devices with the physical assistance (often at the arm, elbow, or wrist) of a facilitator are consistently doubted as the authors of their own messages.

1: Defining a Rhetoric of Touch

1. The use of American Sign Language at this time was generally not supported. Keller was also linked with Alexander Graham Bell, a proponent of the oralism movement, who did not support education based on manual sign language.

2. Keller attempted to understand her life in relation to Sullivan, featuring their partnership in several of the books she wrote in her lifetime. In the disability community, it is worth noting that there is tension over how to interpret Keller's status as a "poster child" and Keller and Sullivan's relationship. Georgina Kleege, for example, expresses frustration over Keller and her legacy. In the film *Vital Signs,* Robert DeFelice also expresses mixed emotions over the story of Keller's life, especially as it intertwined with Sullivan's.

3. Collaborative models of rhetoric and composition have accounted for a dual sense of authorship and *ethos* when a rhetor collaborates with another rhetor or when rhetor and

audience interact, but these models tend to focus on cognitive collaboration (usually among cognitively nondisabled minds), rather than physical collaboration. Similarly, new media rhetorics, although attentive to materiality and collaboration, generally do not yet provide a way of understanding touch as a rhetorically generative strategy. As Michele Knobel and Colin Lankshear point out, new literacies are "more participatory," "collaborative," and "distributed"; they are less "individuated" and "author-centric" than traditional literacies (9). In recent years the concepts of audience, speaker, writer, listener, viewer, reader, text, and media have been thoroughly mixed and remixed, mediated and remediated, but the senses of sight and sound have been explored, whereas touch has remain underexplored.

4. Kennedy elaborates: "The problem with *pistis* is that it can have at least three different meanings in the text: 'proof,' in the sense of logical demonstration; 'proof' as a formal part of a speech; and 'means of persuasion,' including not only logical demonstration but the trustworthiness of character that a speaker can project and the emotion he conveys" (*A New History of Classical Rhetoric* 60).

5. Regarding the potentials or the senses, Aristotle states that "the sensitive power is not an actuality, but is only potential" (*De Anima* 233). In book 2, chapters 4, 5, and 6, he develops this distinction based on potentiality in relation to the senses.

6. As Paterson describes, "the emotional component of feeling complemented by the perceptual component of touching, find its support in Aristotle through both *alloiosis* (alteration, being affected by) and *paschein* as both active and passive, the active undergoing of experience" (19).

7. Richards and Ogden use the metaphor of the triangle to identify the essential elements in the language situation, including the symbol, referent, and thought or reference. Kinneavy offers the communication triangle of encoder, decoder, reality, and signal (language) (*A Theory of Discourse* 18–19). Bitzer uses three elements to explain what constitutes a rhetorical situation: exigency, constraints, and situation. For a comprehensive history on the use of the triad in composition and rhetoric, see Kevin Dvorak, "The Rhetorical Triangle."

8. Garret and Xiao, like many other revisers of the rhetorical situation, focus on "culture's discourse tradition" (30) rather than nondiscursive elements.

9. Debra Hawhee's *Moving Bodies: Kenneth Burke at the Edges of Language* develops this coconstitutive relationship among bodies in Burkean thought.

10. More specifically, "the *killing* of something is the *changing* of it, and the statement of the thing's nature before and after the change is an *identifying* of it" (Burke, *A Rhetoric of Motives* 20).

11. Like Welch, M. Jimmie Killingsworth notes that "the use of the term *proof* runs counter to modern usage—you aren't supposed to 'invent' proof in an argument—a difference that probably accounts for the tendency in modern rhetoric and composition to substitute the term *appeal* not only in textbooks, where the term prevails above all others, but in many scholarly sources" (250). Killingsworth also notes that Kennedy uses the term "proof" in the notes of his translation but uses the Greek term *pisteis* in the main text.

12. Diana Fuss writes that Keller would become the "chief cipher for the new sight and sound technologies of the late nineteenth and early twentieth centuries" (134). Keller was pictured with a radio, a phonograph, and a Dictaphone in her lifetime.

13. See Michael Hassett's "Sophisticated Burke: Kenneth Burke as a Neosophistic Rhetorician" for exploration of Burke in relation to the sophists.

2: Locating Touch

1. Identity and identification are not the same. As Ratcliffe explains, "Identification is inextricably linked with identity but does not directly correspond to it" (51). See Fuss (2–3) and Ratcliffe (51–53) for a discussion of these differences in-depth.

2. Few disability theorists have paid sustained attention to the role of nondisabled people in the struggle against ableism. As Tom Shakespeare has recently pointed out, despite "the commendable urge to put disabled people at the centre of analysis, politics and policy has sometimes led to a writing out of non-disabled people" (*Disability Rights and Wrongs* 186).

3. The issues of identity and identity politics are too capacious and complicated to treat comprehensively in this chapter, but Mollow, Clare, Siebers, Linton, and Davis, among others, discuss the issues involved at length. In particular, Siebers in *Disability Theory* (11–15) explains how identity has been understood in limiting ways via the "ideology of ability" (12). For an exploration of identity and difference in dynamic and relational registers in the teaching of writing, see Kerschbaum.

4. In disability studies, a similar dichotomy exists. As Siebers states in *Disability Theory*, embodiment and the body remain an area of contention in disability studies. Many revisions to the social model, which can be interpreted as focusing overly on social construction, include attention to bodily experience. For example, see Shakespeare and Watson, Hughes, and Clare.

5. Many scholars in disability seek to revise a "medical model" approach to disability focused on deficit and impairment with a "social model" approach that includes attention to the larger cultural and social structures that contribute to disability discrimination. The social model, however, has also undergone numerous revisions. For more information on the models and their revisions, see Shakespeare ("The Social Model of Disability" 2006) and Thomas.

6. Burke's invocation of nature positions substance as material and discursive and individual and social: "men are not only in nature. The cultural accretions made possible by the language motive become 'second nature' with them. Here again we confront the ambiguities of substance, since symbolic communication is not merely external instrument but also intrinsic to men as agents" (*A Grammar of Motives* 33). See Ratcliffe's argument against a reading of essentialization regarding Burke's first and second natures (194n11).

7. Nancy's question of "proper" is interpreted by Derrida as a "question of being one" or "one's 'own'" (*On Touching* 97).

8. Some disability studies theorists critique phenomenology for overly individualistic approaches to embodiment (Turner 52), but the texts I explore show Merleau-Ponty and Nancy offering accounts interconnected with a range of bodies and contexts.

9. Deleuze also draws on Foucault to explore folds of the mind. According to Deleuze, Foucault is "haunted by this theme of an inside which is merely the fold of the outside, as if the ship were a folding of the sea" (*Foucault* 97). He uses the example of a Renaissance madman put to sea to illustrate this theme.

10. Like Merleau-Ponty and Nancy, Deleuze is also interested in "doubling," which he explores via the fold (*Foucault* 98).

11. Susan Squier explores Tollifson's relationship to Zen practice more fully in relation to questions of identity in "Meditation, Disability, and Identity."

12. This openness and receptivity aligns, to a certain extent, with Ratcliffe's notion of rhetorical listening: "a stance of openness that a person may choose to assume in relation to any person, text, or culture" (1).

3: Feeling *Logos*

1. George Kennedy remarks that *logos* is the "mode of proof found in the argument and most characteristic of rhetoric" as Aristotle understands it (*Classical Rhetoric & Its Christian and Secular Tradition* 82). A frequent characteristic of *logos* is that is usually connotes logical reasoning and rationality, as I describe more in-depth in the following section. It is worth noting, however, that interpretations of *logos* do not have to support rationality exclusively. Critiques of logocentrism in rhetoric, for example, provide alternatives. Valuable critiques of misplaced emphasis on rationality in *logos* and argument have also been explored by feminist, postmodern, and deconstructive rhetorics. In this chapter, however, I work with the pervasive assumption that *logos* remains associated with a kind of rationality that is beyond the reach of people with psychological disabilities, which has yet to be fully addressed even in revisionist understandings of rhetoric, as Price also points out (*Mad at School* 32).

2. The handshake is featured between the actors playing Lopez and Ayers in the film *The Soloist*, with these resonances and more.

3. W. K. C. Guthrie notes that Empedocles "had a reputation as an orator and teacher of rhetoric . . . and Gorgias, also a Sicilian, was his pupil" (*A History of Greek Philosophy* 135).

4. As Patricia Bizzell and Bruce Herzberg discuss, the traditional reading of Aristotle's classification system in which the major argumentative structures are contained in *logos* "has tended to privilege rational appeals and devalue pathetic and ethical appeals" (173). Alternatively, William Grimaldi has proposed a revision of this classification by dividing artistic proof into enthymeme and example, of which *logos, ethos,* and *pathos* are further divisions. Even revisionist interpretations of the Aristotelian appeals, which focus on "rhetoric as legitimately appealing to the 'whole person,' not just to the 'rational being,'" still assume a psychologically able and reasonable rhetor, as William Grimaldi envisions that "rational, ethical, and pathetical appeals are typically present in all reasonable arguments" (qtd. in Bizzell and Herzberg 175).

5. Guthrie explains, "Although Love and Strife are invisible and unimaginably fine and tenuous (moreso, obviously, than air or fire), and although their influence is in the first place a psychological one, their spiritual character is not yet completely divorced from physical form" (*A History of Greek Philosophy* 159). Simon Trépanier writes, "It is to these lines we owe our clearest account of the six everlasting four principles, the four 'roots' earth, air, fire and water, and the two 'psychological powers' Love and Strife, and of their interaction to form a never-ending cosmic cycle" ("The Structure of Empedocles' Fragment 17").

6. As Guthrie explains, figuring out exactly how Empedocles may have or have not used the actual term *logos* is difficult (*A History of Greek Philosophy* 160–61, 211). Rather than attempt to offer an answer to this question, I focus on how Empedocles performs and models a form of felt *logos*, specifying *logos* in Empedocles's philosophy as operating in relation to the three areas of application that Kerferd describes, which are clearly at work in fragment 17 and others. As Guthrie, as well as other scholars, relate, to a certain extent, even if Empedocles did not use the word *logos*, "he was perfectly capable of explaining without its use" certain aspects of its range of application, such as ratio and proportion (*A History of*

Greek Philosophy 161n1). Scholars such as Trépanier, Wright (164, 206, 209), Enos ("Aristotle, Empedocles, and the Notion of Rhetoric" 13), and Lambridis (88) also use *logos* to elucidate Empedocles's theories.

7. Unless otherwise noted, I am using Empedocles's fragments as arranged by Hermann Diels and Walther Kranz and translated by John Burnet. To indicate this, I provide the arrangement of Diels and Kranz in the form of the number of the "B" fragments (the "A" group is reserved for other sources); then I provide the page number of Burnet's translation. Fragment 17 is a slightly different case, however, because of lines recently discovered by Alain Martin and Oliver Primavesi in the 1990s. In this excerpt, following Trépanier, the first thirty-five lines are from Diels and Kranz, while the remaining lines of the newly reconstructed fragment, the bulk of which I quote later, follow Martin and Primavesi. The first line of fragment 17, however, seems to differ slightly, with Trépanier using "Double is my account" and Diels and Kranz using "I shall tell thee a twofold tale."

8. For a complex and nuanced approach to how Empedocles's *logos* "grows" in somewhat contradictory but generative ways in fragment 17, see the suggestions Trépanier offers in the epilogue to his analysis.

9. The discovery of the Strasbourg papyrus by Martin and Primavesi in the 1990s added an additional thirty-four partial and complete verses to the thirty-five lines of the fragment, making it not only the longest fragment from Empedocles but also the most complete pre-Socratic fragment. Fortuitously these recently discovered lines from Martin and Primavesi can be understood to fit into the existing lines and even complete Empedocles's ideas, although scholarly conversation is still evolving (Inwood). Also see Trépanier for more details on this.

10. As Guthrie details, contemporary readings of Empedocles's theory on the formation of bodies often argue that a pre-Darwinian sense of the survival of the fittest shapes a desire for "whole-natured forms," seemingly imbuing a telos of able-bodiedness and normalcy into Empedocles's theories (31.B.62; 214). This view often cites Simplicius's interpretation of the relationship between Love and Strife: "Empedocles says that during the first rule of Love . . . there came into being at random parts of animals such as heads, hands, and feet, and then there come together those 'oxen with the heads of men' and the converse, and 'as many of these parts were fitted together in such a way as to ensure their preservation and became animals and survived, because they fulfilled mutual needs. . . . All that did not come together according to the proper formula [*logos*] perished" (qtd. in Guthrie, *A History of Greek Philosophy* 204). Robert Garland calls the various living "creatures," including "hybrid monsters" with "deformities," the "evolutionary left-overs" of Empedocles's progressive stages for the generation of living beings (174). Following Aetius, Garland and other commentators establish four progressive stages by which the various deformed body parts join to make fully evolved, whole human bodies. Empedocles, however, does not give any indications on what the "proper" *logos* is for these whole-natured forms, as Garland notes. At most he cites chance as the factor in determining proportions. Even in Aphrodite's "perfect" harbor, for example, "the somewhat random coming together of the roots in it results in the imperfection" that accompanies any form of life (Wright 238). In addition Empedocles spends dozens of lines describing the variously shaped and manifold forms of flesh, disabled bodies included, and qualifies his single mention of "whole-natured" forms: "These did the fire, desirous of reaching its like, send up, showing as yet neither the charming form of the limbs, nor yet the voice and parts that are proper to men" (31.B.62; 215). Even among whole forms,

there is variation and difference, deviating from the typical "charming form." Furthermore, as Guthrie remarks, the contemporary commentators who interpret this sense of *logos* as a precursor of Darwinian theory of the survival of the fittest merely reveal "superficial resemblances . . . reached from very different premises" (204–5).

11. Wright makes this argument by relying on "the painting simile . . . as a general guide," a decision which also supports the importance of the tactile imagery Empedocles employs (209).

12. Wright suggests that the metaphor of Aphrodite's "perfect harbors" is "unexpected" but suspects that "the reference is to her womb, where tissues are first formed" (238). However, considering Empedocles's reliance on tactile metaphors and examples to explain his theory of pores and effluences—from the temple painters to the water clock—it is far from unexpected that Aphrodite is characterized by manual actions in which she brings things together in diverse ways.

4: Habituating *Ethos*

1. As Frith describes, in the 1940s Kanner and Asperger, working separately, studied a similar condition (*Autism and Asperger Syndrome* 5–10). Kanner described his group of eleven boys as existing in their own world and published "Autistic Disturbances of Affective Contact" in 1943. Asperger saw similar "deficits" in the boys he studied but also the potential of "originality of thought and experience. This can often lead to exceptional achievements in later life" (37). His paper was not translated into English until the 1980s. The lines between autism and Asperger's syndrome can be blurry. In fact the revised *Diagnostic and Statistical Manual of Mental Disorders V* subsumes the diagnostic criteria of Asperger's syndrome under autism spectrum disorders.

2. In the autistic community and beyond, people without autism are often referred to as "neurotypicals." People on the spectrum may use the term "neuroatypical," "autistic," or "person with autism" to refer to themselves, although identity-first language is often preferred to person-first language. This terminology is generally consistent with an approach to cognitive difference based on neurodiversity, which is an appreciation of the wide range of neurologies and cognitive styles of people both on and off the autism spectrum. For a sketch of the history of the neurodiversity movement, see Savarese and Savarese, "The Superior Half of Speaking."

3. Regarding the implications of these assumptions for the writing classroom, see Jurecic's portrayal of her student Gregory, whom she identifies as having Asperger's syndrome and describes as being unable to write for an audience (435). For spirited critiques of the "normate stance" in this analysis, see Lewiecki-Wilson and Dolmage as well as Heilker. Also see Yergeau's "Aut(hored)ism" and "Circle Wars."

4. More broadly, Aristotle eliminates people with disabilities from the ranks of rhetors and even humanity, as James C. Wilson and Cynthia Lewiecki-Wilson have explored (13).

5. Critics of facilitated communication, skeptical of the function of touch in the system of physical support, argue that a "Clever Hans" or "Ouija board" effect, often unintentional, via the facilitator moves the communicators' hands, arms, or fingers toward certain letters or words to create messages. Studies of the authenticity of facilitated communication show conflicting results, but I take the approach of "assumed competence" in the study of communication by people with autism (Biklen). For more information, see Biklen's *Communication*

Unbound and Biklen et al.'s *Autism and the Myth of the Person Alone*; for an overview of the controversies, shortcomings, and ultimate potentials of FC, see Savarese's *Reasonable People*.

6. Blackman and other people with autism often experience problems with proprioception. Sometimes called the sixth sense, proprioception is a sense of the body's positioning and movement in relation to itself and other things, environments, and bodies.

7. Detienne and Vernant author the only current book-length study of *mētis*; citations about *mētis* in this section, unless otherwise specified, are from their work *Cunning Intelligence in Greek Culture and Society*.

8. Hawhee focuses on the various bodies of beings endowed with *mētis*, arguing for *mētis* as a bodily intelligence and exploring the movements, twists, and turns of bodies that exhibit *mētis*, including the octopus, the sea eel, and the wrestler. This embodied *mētis* focuses primarily on able bodies, such as those of the athlete and the wrestler. Also emphasizing *mētis* as a body rhetoric, Jay Dolmage applies *mētis* to disability, arguing that "we need to recognize *mētis* as a rhetoric, thus recognizing the body as rhetorical and thereby valorizing our own and our students' bodily differences as meaningful and meaning-making" ("Breathe upon Us an Even Flame" 119). More recently Dolmage has explored *mētis* in relation to rhetorical histories and theories focused on female bodies ("Metis, *Mētis*, Mestiza, Medusa").

9. Detienne and Vernant elaborate, writing that the same adjective used to describe the octopus is also used to describe the "monster Typhon, too, [who] is *polúplokos*: a multiple creature 'with a hundred heads' whose trunk tapers out into its eel-like limbs" (37).

10. Detienne and Vernant draw from Aristophanes and Plato's *Laws* and *Politics* for these meanings (53–54).

11. As Detienne and Vernant relate, "The daughter of Zeus and *Mētis* made her appearance amid a burst of light and tumult: 'shining in the brilliance of her arms, a dazzling vision of bronze,' she came into the world emitting a great war cry" (181). Furthermore, "It is through her mother that Athena is well-endowed in *mētis*," and it is "because she is the fruit of *Mētis*' womb that she herself is sometimes called *Mētis*, just as her mother was" (179).

12. In Hesiod's *Works and Days*, "it is the 'handman of Athena' who alone is competent to make the farmer's plough, to 'embed the piece of curved wood (*gúēs*) in the heel which carries the ploughshare and then fix it carefully to the beam'" (Detienne and Vernant 179). She is also known for her abilities of spinning and weaving, along with weaving and twisting, and Athena's facilitative actions of "adjusting" and "binding" are part of the manual skill of *mētis* (184n6).

13. Athena's bit is distinctive; even though it could be claimed by Hephaestus's blacksmithing *mētis*, Detienne and Vernant note that Pindar is "quite specific on this point," writing that the "clue to the mode of operation peculiar to Athena lies in the mythical representation of this instrument: she is the deity who presents to man, in the form of an instrument, a power both technical and magical" (197). For more on Athena's *mētis* and its implications for cross-species interaction, see Walters's "Animal Athena."

14. See Walters's "Autistic *Ethos* at Work" for more information on how Grandin and Prince-Hughes craft an *ethos* in the specific contexts of professional and technical communication. This article is a part of the valuable special issue of *Disability Studies Quarterly* that offers insightful connections between autism and rhetoric.

15. It is worth noting that Grandin and Prince-Hughes can be understood as atypical examples of people with autism. In the autistic community, people such as Grandin and, to a lesser extent, Prince-Hughes are sometimes referred to as "shiny autistics," in that each denotes "an individual who is held up as an example of 'what autistics should be.' . . . Most shiny autistics . . . are high-functioning individuals who become famous by writing a book or otherwise sharing their experience as an autistic. It is felt by other autistics that society at large generalizes the lives of the shiny autistics and expects all autistics to be the same or very similar" ("Shiny Autistic").

16. Grandin elaborates on one hypothesis regarding the relationship between touch and autism: "Because of sensory dysfunction, autistic children crave added tactile stimulation. They prefer (proximal) sensory stimulation such as touching, tasting and smelling as opposed to the distant (distal) sensory stimulation of hearing or seeing. In the developing nervous system the proximal senses develop first. In birds and mammals the tactile senses develop first. This may explain why a child with a damaged or immature nervous system prefers the proximal senses" (*Emergence* 37).

17. Grandin's animal handling and slaughtering systems can also be clearly associated with the tricking and trapping elements of *mētis*, although, as Grandin explains, her machines are designed to be as humane as possible (*Thinking in Pictures* 153–55).

5: Grasping *Pathos*

1. Visuality and emotion are particularly linked in contexts of disability. Garland-Thomson has explored the visual rhetorics of disability in popular photography, focusing on emotions such as "the wondrous, the sentimental, the exotic, and the realistic" ("The Politics of Staring" 58). As Garland-Thomson suggests, mediums such as print and photography, along with other visual media, may reinforce traditional cultural scripts. Visual media, for example, authorize staring, which "creates an awkward partnership that estranges and discomfits both viewer and viewed" (57). With photography, staring imposes limits, "eliminating the possibility for interaction or spontaneity between viewer and viewed" (58). Similarly, as Couser explains in "Signifying Bodies," emotions based on rhetorics of triumph, nostalgia, and spiritual compensation dominate the cultural script of disability autobiography, reinforcing limited emotions surrounding disability.

2. Other definitions emphasize *pathos* as a "quality that arouses pity and sorrow" (*OED*).

3. Gorgias also believes that emotional appeals are beneath logical and ethical arguments, explaining to his audience, "Appeals to pity and entreaties and the intercession of friends are useful when a trial takes place before a mob, but among you, the first of the Greeks and men of repute, it is not right to persuade you with the help of friends or entreaties or appeals to pity" (*Defense of Palamedes*, trans. Kennedy 62). In this view, the "best" audiences are not swayed by emotion.

4. Furthermore according to Aristotle, "Old men may feel pity, as well as young men, but not for the same reason. Young men feel it out of kindness; old men out of weakness, imagining that anything that befalls any one else might easily happen to them" (*Rhetoric*, trans. Roberts 1390a19–22; 86).

5. For Aristotle, a similar relationship of proximity and distance directs the emotion of pity, which is related to fear: "We feel pity whenever we are in the condition of remembering

that similar misfortunes have happened to us or ours, or expecting them to happen in the future" (*Rhetoric,* trans. Roberts 1386a1–4; 77).

6. In this way *kairos* is similar to *mētis* and may in fact be a type of *mētis.* As Detienne and Vernant suggest, *kairos* was introduced after *mētis,* with both terms meaning navigation or looking ahead and seeing a propitious moment for steering a vehicle such as a ship (224–25). Hawhee too relates the two concepts, calling *mētis* the "mode" of actors such as sophists and athletes with *kairos* operating as the timing of that mode (*Bodily Arts* 66).

7. As Hawhee notes, Pindar applies this meaning of *kairos* to discourse, advising, "If you should speak to the point by combining the strands of many things in few words, less criticism follows from men" (*Pythian* 1.82; qtd. in Hawhee, *Bodily Arts* 67). As Hawhee explains, the more "tightly woven" a speech, the "fewer openings or opportunities listeners will have to refute" (67).

8. This instance also illustrates the close relationship between *kairos* and *mētis.* As Jay Dolmage explains, "*Kairos,* the idea of invention only within shifting contexts, only in the world of *tuchē*—of the winds of chance—demands *mētis,* a way to be even more mobile, polymorphic, and cunning than the world itself" ("Breathe Upon Us an Even Flame" 121). Hockenberry is acting with *mētis,* especially by moving among different human, animal, and technical bodies, as much as with *kairos.*

6: Teaching Touch, Touching Technology

1. These equal quantities of air and blood recall the proportions of *logos* Empedocles used to describe Hephaestus's making of blood and bone, a process by which disability was built into the construction of a diverse range of bodies, "manifold forms of flesh" (31.B.98; 219).

2. See Enos ("Inventional Constraints on the Technographers of Ancient Athens") and Consigny (*Gorgias, Sophist and Artist*) for other examples of how the water clock was used, especially in relation to the timing of speech-making situations.

3. See chapter 3 for these tactile examples.

4. Zielinski elaborates, "At first glance, it may appear somewhat redundant in the age of unlimited reproducibility of things and organisms to study the ideas of a philosopher who formulated his doctrine in fine hexametric poetry two-and-a-half-thousand years ago" (41–42). Yet Zielinski characterizes the continued relevance of Empedocles's theory of the interface to people and machines by writing that "because it is perfect, building it will never be possible. However, precisely because it possesses this potential impossibility, the theory is entirely worthy of consideration for dealing with existing interfaces, which purport to have already established compatibility between the one and the other" (55).

5. Paterson also reinforces the relevance of Aristotelian theories for haptics, noting that the "themes of flesh, proximity, vulnerability and contact raised by Aristotle remain conceptually pertinent" in current media studies (158).

6. The shadow bodies wearing the white earbuds in iPod commercials, although animated in various colors and rhythms, are also always able bodies.

7. MacNealy considers teacher-research as growing out of criticisms of traditional research. Teacher-researchers assume that "subjective data is valuable, that the research should be carried out as naturalistically as possible in real life situations, and that data collection and analysis must be a collaborative rather than hierarchical effort so that the

distinction between researcher and research subjects is blurred, if not obliterated" (233). Furthermore teacher-research is often "action research, undertaken with the idea that change will occur in the researcher as well as the research subjects" (233). Objectives of a teacher-researcher methodology align with the aims of disability studies methodologies. Like teacher-researcher methods, disability studies methods are critical of traditional research and seek to uncover and value the subjective experience of disability. Linton, for example, in "claiming" disability, argues for valuing subjective experiences of identity and knowledge, and Geof Mercer describes "lay accounts" and "information expressed in feelings, attitudes and actions" as integral to disability research (231).

8. In a different classroom study about accessible technologies in the writing class, I experienced a group of students in which the majority identified as nondisabled. For details on that negotiation and more information on disability studies and teacher-researcher methodologies, see Walters, "Toward an Accessible Pedagogy." I also consulted Price's "Disability Studies Methodology" and Vidali's "Performing the Rhetorical Freak Show" regarding issues such as this surrounding this study.

9. Following review, the study qualified for exemption status because it fell under the realm of normal educational practices and settings; however, I still made students aware that their participation was voluntary and would be confidential. To ensure confidentiality, I have not used students' names.

10. See Hillis, *Digital Sensations;* Sheridan, "Human and Machine Haptics in Historical Perspective"; and Stone, "Haptic Feedback" for more detailed histories of haptics.

11. Search "iPad" and "cat," "dog," or "toddler" in Youtube, and dozens of videos surface.

Bibliography

Ackerman, Diane. *A Natural History of the Senses*. New York: Vintage Books, 1991.

Alcoff, Linda Martin. *Visible Identities: Race, Gender, and the Self*. Oxford: Oxford University Press, 2006.

Anzaldúa, Gloria. *Borderlands/La Frontera: The New Mestiza*. San Francisco: Aunt Lute, 2007.

"Apps for Autism: Communicating on the iPad." *60 Minutes*. CBS News. http://www.cbsnews.com/8301-18560_162-57460553/apps-for-autism-communicating-on-the-ipad/ (accessed January 29, 2013).

Aristotle. *De Anima*. Trans. Kenelm Foster and Silvester Humphries. New Haven, Conn.: Yale University Press, 1951.

——. *Nicomachean Ethics*. Trans. David Ross. Ed. Lesley Brown. Oxford: Oxford University Press, 2009.

——. *On Rhetoric: A Theory of Civil Discourse*. Trans. George A. Kennedy. Oxford: Oxford University Press, 1991.

——. *Rhetoric*. Trans. W. Rhys Roberts. Mineola, N.Y.: Dover, 2004.

Arnold, Matthew. "Empedocles on Etna." In *The Poems of Matthew Arnold*, 80–112. Boston: Elibron Classics Adamant Media, 2006.

——. "Sohrab and Rustum and Other Poems." In *The Poems of Matthew Arnold*, 182–205. Boston: Elibron Classics Adamant Media, 2006.

Asperger, Hans. "'Autistic Psychopathy' in Childhood." In *Autism and Asperger Syndrome*, ed. Uta Frith, 37–92. Cambridge: Cambridge University Press, 1991.

Atwill, Janet M. *Rhetoric Reclaimed: Aristotle and the Liberal Arts Tradition*. Ithaca, N.Y.: Cornell University Press, 1998.

Barber-Fendly, Kimber, and Chris Hamel. "A New Visibility: An Argument for Alternative Writing Programs for Students with Learning Disabilities." *College Composition and Communication* 55.3 (2004): 504–35.

Barnartt, Sharon, and Richard Scotch. *Disability Protests: Contentious Politics 1970–1999*. Washington, D.C.: Gallaudet University Press, 2001.

Baron-Cohen, Simon. *Mindblindness: An Essay on Autism and Theory of Mind*. Cambridge, Mass.: MIT Press, 1995.

Bartholomae, David, and Anthony Petrosky. *Ways of Reading Words and Images*. Boston: Bedford/St. Martin's, 2003.

Barton, Ellen L. "Textual Practices of Erasure: Representations of Disability and the Founding of the United Way." In *Embodied Rhetorics: Disability in Language and Culture*, ed. James C. Wilson and Cynthia Lewiecki-Wilson, 169–99. Carbondale: Southern Illinois University Press, 2001.

Baumlin, James S. "Introduction: Positioning *Ethos* in Historical and Comtemporary Theory." In *Ethos: New Essays in Rhetorical and Critical Theory*, ed. James S. Baumlin and Tita French Baumlin, xi–xxxi. Dallas: Southern Methodist University Press, 1994.

Baumlin, James S., and Tita French Baumlin. "Psyche/Logos: Mapping the Terrain of Mind and Rhetoric." *College English* 51.3 (1989): 245–61.

——, eds. *Ethos: New Essays in Rhetorical and Critical Theory.* Dallas: Southern Methodist University Press, 1994.

Berlin, James A. *Rhetorics, Poetics, Cultures: Refiguring College English Studies.* Urbana, Ill.: NCTE, 1996.

Bérubé, Michael. "Foreword: Pressing the Claim." In Simi Linton, *Claiming Disability: Knowledge and Identity*, vii–xi. New York: New York University Press, 1998.

Biesecker, Barbara A. "Coming to Terms with Recent Attempts to Write Women into the History of Rhetoric." *Philosophy & Rhetoric* 25.2 (1992): 140–61.

——. "Rethinking the Rhetorical Situation from within the Thematic of 'Différance.'" *Philosophy & Rhetoric* 22.2 (1989): 110–30.

Biklen, Douglas. *Communication Unbound: How Facilitated Communication Is Challenging Traditional Views of Autism and Ability/Disability.* New York: Teacher College Press, 1993.

Biklen, Douglas, with Richard Attfield, Larry Bissonnette, Lucy Blackman, Jamie Burke, Alberto Frugone, Tito Rajarshi Mukhopadhyay, and Sue Rubin. *Autism and the Myth of the Person Alone.* New York: New York University Press, 2005.

Bitzer, Lloyd F. "The Rhetorical Situation." *Philosophy & Rhetoric* 1.1 (1968): 1–14.

Bizzell, Patricia, and Bruce Herzberg, eds. *The Rhetorical Tradition: Readings from Classical Times to the Present.* 2nd ed. Boston: Bedford/St. Martin's, 2001.

Blackman, Lucy. *Lucy's Story: Autism and Other Adventures.* London: Jessica Kingsley, 1999.

Bolter, Jay David. *Turing's Man: Western Culture in the Computer Age.* Chapel Hill: University of North Carolina Press, 1984.

Bolter, Jay David, and Richard Grusin. *Remediation: Understanding New Media.* Cambridge, Mass.: MIT Press, 2000.

Booth, Wayne C. "The Rhetorical Stance." *College Composition and Communication* 14.3 (1963): 139–45.

Breslin Oral History. "Disability Rights and Independent Living." Bancroft Library. University of California, Berkeley.

Brueggemann, Brenda Jo. *Lend Me Your Ear: Rhetorical Constructions of Deafness.* Washington, D.C.: Gallaudet University Press, 1999.

Brueggemann, Brenda Jo, and James A. Fredal. "Studying Disability Rhetorically." *Disability Studies Quarterly* 17.4 (1997): 251–57.

Brueggemann, Brenda Jo, Linda Feldmeier White, Patricia A. Dunn, Barbara A. Heifferon, and Johnson Cheu. "Becoming Visible: Lessons in Disability." *College Composition and Communication* 52.3 (2001): 368–98.

Brueggemann, Brenda Jo, and Cynthia Lewiecki-Wilson, eds., with Jay Dolmage. *Disability and the Teaching of Writing: A Critical Sourcebook.* Boston: Bedford/St. Martin's, 2007.

Burke, Kenneth. *Attitudes toward History.* 3rd ed. Berkeley: University of California Press, 1984.

——. *A Grammar of Motives.* 1945. Reprint, Berkeley: University of California Press, 1969.

——. *Language as Symbolic Action: Essays on Life, Literature, and Method.* Berkeley: University of California Press, 1966.

——. *The Philosophy of Literary Form: Studies in Symbolic Action.* 1941. Reprint, Berkeley: University of California Press, 1973.

——. *A Rhetoric of Motives.* 1950. Reprint, Berkeley: University of California Press, 1969.

Burnet, John, trans. "Empedokles of Akragas." In *Early Greek Philosophy,* ed. John Burnet, 4th ed, 197–250. New York: Meridian Books, 1957.

Bush, Vannevar. *Endless Horizons.* Washington, D.C.: Public Affairs Press, 1946.

Butler, Judith. *Bodies That Matter: On the Discursive Limits of "Sex."* New York: Routledge, 1993.

Buxton, Bill. "Multitouch Systems I Have Known and Loved." http://www.billbuxton.com/multitouchOverview.html (accessed January 24, 2013).

Challis, Ben P., and Alistair D. N. Edwards. "Design Principles for Tactile Interaction." In *Haptic Human-Computer Interaction: First International Workshop Glasgow, UK, August/September 2000 Proceedings,* ed. Stephen Brewster and Roderick Murray-Smith, 17–24. Berlin: Springer, 2001.

Charlton, James I. *Nothing about Us without Us: Disability Oppression and Empowerment.* Berkeley: University of California Press, 1998.

Chrétien, Jean-Louis. *The Call and the Response.* Trans. Anne A. Davenport. New York: Fordham University Press, 2004.

Chrisman, Wendy L. "A Reflection on Inspiration: A Recuperative Call for Emotion in Disability Studies." *Journal of Literary & Cultural Disability Studies* 5.2 (2011): 173–84.

Cicero. *De Oratore.* Trans. E. W. Sutton and H. Rackham. Cambridge, Mass.: Harvard University Press, 1942.

Cixous, Hélène. "The Laugh of the Medusa." Trans. Keith Cohen and Paula Cohen. *Signs* 1.4 (1976): 875–93.

Clare, Eli. *Exile and Pride: Disability, Queerness, and Liberation.* Cambridge: South End Press, 1999.

Classen, Constance. "Fingerprints: Writing about Touch." In *The Book of Touch,* ed. Constance Classen, 1–9. Oxford: Berg, 2005.

——. *The Deepest Sense: A Cultural History of Touch.* Urbana: University of Illinois Press, 2012.

——, ed. *The Book of Touch.* Oxford: Berg, 2005.

Cochran-Smith, Marilyn, and Susan L. Lytle. *Inside/Outside: Teacher Research and Knowledge.* New York: Teachers College Press, 1993.

Condillac, Étienne Bonnot de. *Treatise on the Sensations.* Trans. Geraldine Carr. Los Angeles: University of Southern California School of Philosophy, 1930.

——. *Treatise on the Sensations.* Trans. Franklin Philip, with the collaboration of Harlan Lane. Hillsdale, N.J.: Lawrence Erlbaum Associates, 1982.

Consigny, Scott. *Gorgias, Sophist and Artist.* Columbia: University of South Carolina Press, 2001.

——. "Rhetoric and Its Situations." *Philosophy and Rhetoric* 7.3 (1974): 175–86.

Cope, Bill, and Mary Kalantzis, eds. *Multiliteracies: Literacy Learning and the Design of Social Futures.* London: Routledge, 2000.

Corbett, Edward P. J. "The Rhetoric of the Open Hand and the Rhetoric of the Closed Fist." *College Composition and Communication* 20.5 (1969): 288–96.

Corker, Mairian, and Sally French. "Reclaiming Discourse in Disability Studies." In *Disability Discourse,* ed. Mairian Corker and Sally French, 1–11. Philadelphia: Open University Press, 1999.

——, eds. *Disability Discourse.* Philadelphia: Open University Press, 1999.

Corker, Mairian, and Tom Shakespeare. *Disability/Postmodernity: Embodying Disability Theory.* New York: Continuum Press, 2002.

Couser, G. Thomas. "Signifying Bodies: Life Writing and Disability Studies." In *Disability Studies: Enabling the Humanities,* ed. Sharon Snyder, Brenda Jo Brueggemann, and Rosemarie Garland-Thomson, 109–17. New York: MLA, 2002.

——. *Vulnerable Subjects: Ethics and Life Writing.* Ithaca, N.Y.: Cornell University Press, 2004.

Cressman, Jodi. "Helen Keller and the Mind's Eyewitness." *Western Humanities Review* 54.2 (2000): 108–23.

Crowley, Sharon. "Afterword: The Material of Rhetoric." In *Rhetorical Bodies,* ed. Jack Selzer and Sharon Crowley, 326–64. Madison: University of Wisconsin Press, 1999.

——. "Body Studies in Rhetoric and Composition." In *Rhetoric and Composition as Intellectual Work,* ed. Gary Olson, 177–87. Carbondale: Southern Illinois University Press, 2002.

Crowley, Sharon, and Debra Hawhee. *Ancient Rhetorics for Contemporary Students.* 3rd ed. New York: Pearson Longman, 2004.

"Darius." In *Aquamarine Blue 5: Personal Stories of College Students with Autism,* ed. Dawn Prince-Hughes, 9–42. Athens, Ohio: Swallow Press, 2002.

Davenport, Anne A. "Aristotle and Descartes on Touch." *New Arcadia Review* 2 (2004). http://www.bc.edu/publications/newarcadia/archives/2/aristotledescartes/ (accessed January 29, 2013).

Davis, Lennard J. *Bending Over Backwards: Disability, Dismodernism, and Other Difficult Positions.* New York: New York University Press, 2002.

——. *Enforcing Normalcy: Disability, Deafness, and the Body.* New York: Verso, 1995.

——. "Introduction." In *The Disability Studies Reader,* 2nd ed., ed. Lennard Davis, xv–xviii. New York: Routledge, 2006.

——. "Preface to the Second Edition." In *The Disability Studies Reader,* 2nd ed., ed. Lennard Davis, xiii–xiv. New York: Routledge, 2006.

——, ed. *The Disability Studies Reader.* 2nd ed. New York: Routledge, 2006.

Deleuze, Gilles. *The Fold: Leibniz and the Baroque.* Trans. Tom Conley. Minneapolis: University of Minnesota Press, 1993.

——. *Foucault.* Trans. and ed. Sean Hand. London: Athlone Press, 1988.

Deleuze, Gilles, and Felix Guattari. *A Thousand Plateaus: Capitalism and Schizophrenia.* Trans. Brian Massumi. Minneapolis: University of Minnesota Press, 1987.

"Delusions, Hallucinations and Beliefs: Bugs." Schizophrenia.com. http://www.schizophrenia.com:8080/jiveforums/thread.jspa?threadID=7223 (accessed January 27, 2013).

Derrida, Jacques. *On Touching—Jean-Luc Nancy.* Trans. Christine Irizarry. Stanford, Calif.: Stanford University Press, 2005.

——. "Plato's Pharmacy." In *Dissemination,* trans. Barbara Johnson, 63–171. Chicago: University of Chicago Press, 1981.

Detienne, Marcel, and Jean-Pierre Vernant. *Cunning Intelligence in Greek Culture and Society.* Trans. Janet Loyd. Atlantic Highlands, N.J.: Humanities Press, 1978.

Diagnostic and Statistical Manual of Mental Disorders. 4th ed. Washington, D.C.: American Psychiatric Association, 1994.

Diagnostic and Statistical Manual of Mental Disorders. 5th. ed. Washington, D.C.: American Psychiatric Association, 2013.

Diaz, Jesus. "Apple's iPad is the Future. This Is Why." MSNBC. http://www.msnbc.msn.com/id/36158687/ns/technology_and_science-tech_and_gadgets (accessed January 29, 2013).

Diels, Hermann, and Walther Kranz. *Die Fragmente de Vorsokratiker*. 5th ed. Berlin: Weidmannsche Verlagsbuchhandlung, 1934.

Diogenes Laertius. *Lives of Eminent Philosophers*. Vol. 2. Ed. R. D. Hicks. Cambridge, Mass.: Harvard University Press, 1972.

Disability Rights Education and Defense Fund. "504 Sit-in 20th Anniversary." http://www.dredf.org/504site/504home.html (accessed January 28, 2013).

Dolmage, Jay. "'Breathe upon Us an Even Flame': Hephaestus, History and the Body of Rhetoric." *Rhetoric Review* 25.2 (2006): 119–40.

——. "Metis, *Mētis*, Mestiza, Medusa: Rhetorical Bodies across Rhetorical Traditions." *Rhetoric Review* 28.1 (2009): 1–28.

Dolmage, Jay, and Cynthia Lewiecki-Wilson. "Refiguring Rhetorica: Linking Feminist Rhetoric and Disability Studies." In *Rhetorica in Motion: Feminist Rhetorical Methods and Methodologies*, ed. Eileen Schell and Kelly Rawson, 36–60. Pittsburgh: Pittsburgh University Press, 2010.

Donaldson, Elizabeth J., and Catherine Prendergast. "Introduction: Disability and Emotion: There's No Crying in Disability Studies!" *Journal of Literary & Cultural Disability Studies* 5.2 (2011): 129–36.

Donnellan, Ann M., David A. Hill, and Martha R. Leary. "Rethinking Autism: Implications of Sensory and Movement Differences." *Disability Studies Quarterly* 30.1 (2010). http://dsq-sds.org/article/view/1060/1225 (accessed January 29, 2013).

Duffy, John, with Rebecca Dorner. "The Pathos of 'Mindblindness': Autism, Science and Sadness in 'Theory of Mind' Narratives." *Journal of Literary & Cultural Disability Studies* 5.2 (2011): 201–15.

Dunbar, Robin. *Grooming, Gossip, and the Evolution of Language*. Cambridge, Mass.: Harvard University Press, 1998.

Dunn, Patricia A. *Learning Re-Abled: The Learning Disability Controversy and Composition Studies*. Portsmouth, N.H.: Boynton/Cook Heinemann, 1995.

——. *Talking Sketching Moving: Multiple Literacies in the Teaching of Writing*. Portsmouth, N.H.: Boynton/Cook Heinemann, 2001.

Dvorak, Kevin. "The Rhetorical Triangle—Writer, Reader, Text, and Context/Purpose—in Composition-Rhetoric: A History." Ph.D. diss., Indiana University of Pennsylvania, 2006.

Edbauer, Jenny. "Unframing Models of Public Distribution: From Rhetorical Situation to Rhetorical Ecologies." *Rhetoric Society Quarterly* 35.4 (2005): 5–24.

Ede, Lisa, and Andrea A. Lunsford. "Collaboration and Concepts of Authorship." *PMLA* 116.2 (2001): 354–69.

Enos, Richard Leo. "Aristotle, Empedocles, and the Notion of Rhetoric." In *In Search of Justice: The Indiana Tradition in Speech Communication*, ed. Richard J. Jensen and John C. Hammerback, 5–21. Amsterdam: Rodopi, 1987.

——. "Inventional Constraints on the Technographers of Ancient Athens: A Study of Kairos." In *Rhetoric and Kairos*, ed. Philip Sipiora and James Baumlin, 77–88. Albany: State University of New York Press, 2002.

Fagles, Robert. *The Iliad*. New York: Penguin Books, 1990.

Field, Tiffany. *Touch*. Cambridge, Mass.: MIT Press, 2003.

Finger, Anne. "Writing Disabled Lives: Beyond the Singular." *PMLA* 120.2 (2005): 610–15.

Finnegan, Ruth. "Tactile Communication." In *The Book of Touch*, ed. Constance Classen, 18–25. Oxford: Berg, 2007.

"504 Demonstrations." *CBS Evening News*. April 5, 1977. http://www.youtube.com/watch?v=pbfNJpFni-E (accessed March 22, 2013).

Fleckenstein, Kristie S. "Words Made Flesh: Fusing Imagery and Language in a Polymorphic Literacy." *College English* 66.6 (2004): 612–31.

Foucault, Michel. *The Archeology of Knowledge and the Discourse on Language*. Trans. A. M. Sheridan Smith. New York: Pantheon, 1972.

Freeland, Cynthia. "Aristotle on the Sense of Touch." In *Essays on Aristotle's De Anima*, ed. Martha C. Nussbaum and Amélie Oksenberg Rorty, 227–48. Oxford: Clarendon, 1992.

Freeman, Kathleen, trans. *Ancilla to The Pre-Socratic Philosophers: A complete translation of the Fragments in Diels, Fragmente de Vorsokratiker*. Cambridge: Harvard University Press, 1957.

Fries, Kenny. "Excavation." In *Anesthesia: Poems*. Louisville: The Avocado Press, 1996.

——. "Introduction." In *Staring Back: The Disability Experience from the Inside Out*, ed. Kenny Fries, 1–10. New York: Plume Books, 1997.

——, ed. *Staring Back: The Disability Experience from the Inside Out*. New York: Plume Books, 1997.

Frith, Uta. *Autism: Explaining the Enigma*. Oxford: Basil Blackwell, 1989.

——, ed. *Autism and Asperger Syndrome*. Cambridge: Cambridge University Press, 1991.

Fuss, Diana. *Identification Papers*. New York: Routledge, 1995.

Garland, Robert. *The Eye of the Beholder: Deformity and Disability in the Graeco-Roman World*. Ithaca, N.Y.: Cornell University Press, 1995.

Garland-Thomson, Rosemarie. *Extraordinary Bodies: Figuring Disability in American Culture and Literature*. New York: Columbia University Press, 1996.

——. "The Politics of Staring: Visual Rhetorics of Disability in Popular Photography." In *Disability Studies: Enabling the Humanities*, ed. Sharon Snyder, Brenda Jo Brueggemann, and Rosemarie Garland-Thomson, 56–75. New York: MLA, 2002.

——. *Staring: How We Look*. Oxford: Oxford University Press, 2009.

Garret, Mary, and Xiaosui Xiao. "'The Rhetorical Situation' Revisited." *Rhetoric Society Quarterly* 23.2 (1993): 30–40.

George, Diana. "From Analysis to Design: Visual Communication in the Teaching of Writing." *College Composition and Communication* 52.1 (2002): 11–39.

Gill, Carol J. "A Psychological View of Disability Culture." *Disability Studies Quarterly* 15.4 (1995): 16–19.

Gilyard, Keith. "African American Contributions to Composition Studies." *College Composition and Communication* 50.4 (1999): 626–44.

Glenn, Cheryl. *Unspoken: A Rhetoric of Silence*. Carbondale: Southern Illinois University Press, 2004.

Gorgias. *Defense of Palamedes*. Trans. George A. Kennedy. In *The Older Sophists*, ed. Rosamond Kent Sprague, 54–63. Indianapolis: Hackett, 1972.

——. *Encomium of Helen*. Trans. George A. Kennedy. In *The Older Sophists*, ed. Rosamond Kent Sprague, 50–54. Indianapolis: Hackett, 1972.

Grandin, Temple. *Thinking in Pictures: And Other Reports from My Life with Autism.* New York: Vintage, 1996.

Grandin, Temple, and Margaret M. Scariano. *Emergence: Labeled Autistic.* Novato, Calif.: Arena Press, 1986.

Grimaldi, William M. A. *Studies in the Philosophy of Aristotle's Rhetoric.* Wiesbaden: Franz Steiner Verlag, 1972.

Grosz, Elizabeth. *Volatile Bodies: Toward a Corporeal Feminism.* Bloomington: Indiana University Press, 1994.

Guthrie, W. K. C. *A History of Greek Philosophy.* Vol. 2. Cambridge: Cambridge University Press, 1965.

——. *Myth and Reason.* London: London School of Economics and Political Science, 1953.

Halloran, S. Michael. "Aristotle's Concept of Ethos, or if Not His Somebody Else's." *Rhetoric Review* 1.1 (1982): 58–63.

Hamlyn, D. W. "Aristotle's Account of Aesthesis in the *De Anima.*" *Classical Quarterly* 9.1 (1959): 6–16.

"Handicapped Win Demands: End HEW Occupation." *Black Panther,* May 7, 1977, 6.

Haraway, Donna. *The Companion Species Manifesto: Dogs, People, and Signification Otherness.* Chicago: Prickly Paradigm Press, 2003.

——. *When Species Meet.* Minneapolis: University of Minnesota Press, 2008.

Hassett, Michael. "Sophisticated Burke: Kenneth Burke as a Neosophistic Rhetorician." *Rhetoric Review* 13.2 (1995): 371–90.

Hawhee, Debra. *Bodily Arts: Rhetoric and Athletics in Ancient Greece.* Austin: University of Texas Press, 2004.

——. "Bodily Pedagogies: Rhetoric, Athletics, and the Sophists' Three Rs." *College English* 65.2 (2002): 142–62.

——. *Moving Bodies: Kenneth Burke at the Edges of Language.* Columbia: University of South Carolina Press, 2009.

Heilker, Paul. "Comment/Response: 'Neurodiversity.'" *College English* 70.3 (2008): 319–21.

Heilker, Paul, and Melanie Yergeau. "Autism and Rhetoric." *College English* 73.5 (2011): 485–97

Hilligoss, Susan. *Visual Communication: A Writer's Guide.* New York: Longman, 2000.

Hillis, Ken. *Digital Sensations: Space, Identity, and Embodiment in Virtual Reality.* Minneapolis: University of Minnesota Press, 1999.

Hockenberry, John. "Walking with the Kurds." In *Staring Back: The Disability Experience from the Inside Out,* ed. Kenny Fries, 22–36. New York: Plume Books, 1997.

Howes, David. "Skinscapes: Embodiment, Culture and Environment." In *The Book of Touch,* ed. Constance Classen, 27–39. Oxford: Berg, 2005.

Hughes, Bill. "Disability and the Body." In *Disability Studies Today,* ed. Colin Barnes, Mike Oliver, and Len Barton, 58–97. Boston: Wiley-Blackwell, 2002.

Hunsaker, David M., and Craig R. Smith. "The Nature of Issues: A Constructive Approach to Situational Rhetoric." *Western Journal of Speech Communication* 40.3 (1976): 144–55.

Hyde, Michael J. "Introduction: Rhetorically, We Dwell." In *The Ethos of Rhetoric,* ed. Michael Hyde, xiii–xxviii. Columbia: University of South Carolina Press, 2004.

"I Think I Might Have Schizophrenia." August 23, 2005. http://ehealthforum.com/health/topic40585.html (accessed January 29, 2013).

"I Told You I Wasn't Mad." February 9, 2010. http://www.dailystrength.org/c/Schizophrenia/forum/9062351-i-told-you-i-wasnt (accessed January 30, 2013).

Inwood, Brad, trans. *The Poem of Empedocles.* Rev. ed. Toronto: University of Toronto Press, 2001.

Irigaray, Luce. *An Ethics of Sexual Difference.* Trans. C. Burke and G. C. Gill. Ithaca, N.Y.: Cornell University Press, 1993.

Isocrates. *Antidosis.* Trans. George Norlin. Vol. 2, 181–367. Cambridge, Mass.: Harvard University Press, 1929.

Jablonski, Nina G. *Skin: A Natural History.* Berkeley: University of California Press, 2006.

Jack, Jordynn. "What Are Neurorhetorics?" *Rhetoric Society Quarterly* 40.5 (2010): 405–10.

Jack, Jordynn, and L. Gregory Appelbaum. "This Is Your Brain on Rhetoric: Research Directions for NeuroRhetorics." *Rhetoric Society Quarterly* 40.5 (2010): 411–37.

Jacobs, Dale, and Laura R. Micciche. "Introduction: Rhetoric, Editing, and Emotion." In *A Way to Move: Rhetorics of Emotion & Composition Studies,* ed. Dale Jacobs and Laura Micciche, 1–10. Portsmouth, N.H.: Boynton/Cook Heinemann, 2003.

——, eds. *A Way to Move: Rhetorics of Emotion & Composition Studies.* Portsmouth, N.H.: Boynton/Cook Heinemann, 2003.

Jardin, Xeni. "Review: Apple's iPad is a Touch of Genius." March 31, 2010. http://www.boingboing.net/2010/03/31/a-first-look-at-ipad.html (accessed January 30, 2013).

Jarratt, Susan C. *Rereading the Sophists: Classical Rhetoric Refigured.* Carbondale: Southern Illinois University Press, 1991.

"Jeff Han Demos His Breakthrough Touchscreen." *TED Talks.* February 2006. http://www.ted.com/talks/jeff_han_demos_his_breakthrough_touchscreen.html (accessed January 30, 2013).

"Jim." In *Aquamarine Blue 5: Personal Stories of College Students with Autism,* ed. Dawn Prince-Hughes, 66–75. Athens, Ohio: Swallow Press, 2002.

Jobs, Steve. "iPad Keynote." January 27, 2010. http://www.youtube.com/watch?v=poaUbmdUcCY (accessed January 30, 2013).

Johnson, Harriet McBryde. *Too Late to Die Young: Nearly True Tales from a Life.* New York: Henry Holt, 2005.

——. "Unspeakable Conversations." *New York Times Magazine,* February 16, 2003. http://www.nytimes.com/2003/02/16/magazine/unspeakable-conversations.html?pagewanted=all&src=pm (accessed January 30, 2013).

Johnson, Mary. "Foreword." In *The Ragged Edge,* ed. Barrett Shaw, viii–ix. Louisville, Ky.: Avocado Press, 1994.

Johnson, Robert R. *User-Centered Technology: A Rhetorical Theory for Computers and Other Mundane Artifacts.* Albany: State University of New York Press, 1998.

Johnstone, Christopher Lyle. *Listening to the Logos: Speech and the Coming of Wisdom in Ancient Greece.* Columbia: University of South Carolina Press, 2009.

Jones, Stanley E., and A. Elaine Yarbrough. "A Naturalistic Study of the Meanings of Touch." *Communication Monographs* 52.1 (1985): 19–56.

Jurecic, Ann. "Neurodiversity." *College English* 69.5 (2007): 421–42.

Kanner, Leo. "Autistic Disturbances of Affective Contact." In *Classic Readings in Autism,* ed. Anne M. Donnellan, 11–52. New York: Teachers College Press, 1985.

Keller, Helen. *Midstream: My Later Life.* New York: Greenwood Press, 1968.

——. *The Story of My Life*. New York: Doubleday, 1954.

——. *Teacher, Anne Sullivan Macy*. New York: Doubleday, 1955.

——. *The World I Live In*. Ed. Roger Shattuck. New York: New York Review Books, 2003.

Kendrick, Deborah. "The Sit-in That Ended Segregation of the Disabled." *City Beat*, April 18, 2007. http://www.citybeat.com/cincinnati/article-2503-the-sit-in-that-ended-segregation-of-the-disabled.html (accessed January 30, 2013).

Kennedy, George A. *Classical Rhetoric & Its Christian and Secular Tradition from Ancient to Modern Times*. Chapel Hill: University of North Carolina Press, 1999.

——. *A New History of Classical Rhetoric*. Princeton, N.J.: Princeton University Press, 1994.

Kerferd, G. B. *The Sophistic Movement*. Cambridge: Cambridge University Press, 1981.

Kerschbaum, Stephanie L. "Avoiding the Difference Fixation: Identity Categories, Markers of Difference, and the Teaching of Writing." *College Composition and Communication* 63.4 (2012): 616–44.

Killingsworth, M. Jimmie. "Rhetorical Appeals: A Revision." *Rhetoric Review* 24.3 (2005): 249–63.

Kinneavy, James E. "The Basic Aims of Discourse." *College Composition and Communication* 20.5 (1969): 297–304.

——. *A Theory of Discourse: The Aims of Discourse*. Englewood Cliffs, N.J.: Prentice, 1971.

Kittay, Eva Feder. *Love's Labor: Essays on Women, Equality, and Dependency*. New York: Routledge, 1999.

Kleege, Georgina. *Blind Rage: Letters to Helen Keller*. Washington, D.C.: Gallaudet University Press, 2006.

Knobel, Michele, and Colin Lankshear. "Sampling 'the New' in New Literacies." In *A New Digital Literacies Sampler*, ed. Michele Knobel and Colin Lankshear, 1–24. New York: Peter Lang, 2009.

——, eds. *A New Digital Literacies Sampler*. New York: Peter Lang, 2009.

Korzybski, Alfred. *Science and Sanity: An Introduction to Non-Aristotelian Systems and General Semantics*. 5th ed. Englewood, N.J.: Institute of General Semantics, 1994. http://esgs.free.fr/uk/art/sands.htm (accessed January 30, 2013)

Kress, Gunther. "'English' at the Crossroads: Rethinking Curricula of Communication in the Context of the Turn to the Visual." In *Passions, Pedagogies and 21st Century Technologies*, ed. Gail E. Hawisher and Cynthia L. Selfe, 66–88. Urbana, Ill.: National Council of Teachers of English, 1999.

Krueger, Myron. "Artificial Reality: Past and Future." In *Virtual Reality: Theory, Practice, and Promise*, ed. Sandra K. Helsel and Judith Paris Roth, 19–25. Westport, Conn.: Meckler, 1991.

Kuppers, Petra. "Toward a Rhizomatic Model of Disability: Poetry, Performance, and Touch." *Journal of Literary & Cultural Disability Studies* 3.3 (2009): 221–40.

Kuppers, Petra, and Neil Marcus. *Cripple Poetics: A Love Story*. Ypsilanti, Mich.: Homofactus Press, 2008.

Kuriyama, Shigehisa. *The Expressiveness of the Body and the Divergence of Greek and Chinese Medicine*. New York: Zone Books, 2002.

"Lack of Touch = Depression?" March 3, 2007. http://www.socialanxietysupport.com/forum/f35/lack-of-touch-depression-22131/ (accessed January 30, 2013).

Lambridis, Helle. *Empedocles: A Philosophical Investigation*. Tuscaloosa: University of Alabama Press, 1976.

Lash, Joseph P. *Helen and Teacher: The Story of Helen Keller and Anne Sullivan Macy.* New York: Delacorte Press/Seymour Lawrence, 1980.

Latour, Bruno. *Science in Action: How to Follow Scientists and Engineers through Society.* Cambridge, Mass.: Harvard University Press, 1987.

Lee, Rebecca. "Sympathy through Technology." *ABC News.* July 5, 2007. http://abcnews.go.com/WN/story?id=3348856&page=1 (accessed January 30, 2013).

Leonard, Peter. *Postmodern Welfare: Reconstructing an Emancipatory Project.* London: Sage, 1997.

Lewiecki-Wilson, Cynthia. "Ableist Rhetorics, Nevertheless: Disability and Animal Rights in the Work of Peter Singer and Martha Nussbaum." *JAC: A Journal of Rhetoric, Culture, & Politics* 31.1–2 (2011): 71–101.

——. "Rethinking Rhetoric through Mental Disabilities." *Rhetoric Review* 22.2 (2003): 154–202.

Lewiecki-Wilson, Cynthia, and Jay Dolmage. "Comment/Response: 'Neurodiversity.'" *College English* 70.3 (2008): 314–18.

Lewiecki-Wilson, Cynthia, and Brenda Jo Brueggemann, eds., with Jay Dolmage. *Disability and the Teaching of Writing: A Critical Sourcebook.* Boston: Bedford/St. Martin's, 2008.

Liddell, Henry George, and Robert Scott. *A Greek-English Lexicon.* Oxford: Clarendon, 1996. http://www.perseus.tufts.edu/hopper/text?doc=Perseus:text:1999.04.0057 (accessed January 30, 2013).

Linton, Simi. *Claiming Disability: Knowledge and Identity.* New York: New York University Press, 1998.

——. *My Body Politic: A Memoir.* Ann Arbor: University of Michigan Press, 2007.

Littlefield, Melissa M., and Jenell M. Johnson. *The Neuroscientific Turn: Transdisciplinarity in the Age of the Brain.* Ann Arbor: University of Michigan Press, 2012.

Logan, Shirley Wilson. *"We Are Coming": Nineteenth Century Black Women's Persuasive Discourse.* Carbondale: Southern Illinois University Press, 1999.

Longmore, Paul K. *Why I Burned My Book: And Other Essays on Disability.* Philadelphia: Temple University Press, 2003.

Lopez, Steve. *The Soloist: A Lost Dream, an Unlikely Friendship, and the Redemptive Power of Music.* New York: Berkley Books, 2008.

Lunsford, Andrea A., ed. *Reclaiming Rhetorica: Women in the Rhetorical Tradition.* Pittsburgh: University of Pittsburgh Press, 1995.

Lunsford, Andrea A., and Lisa Ede. "'Among the Audience': On Audience in an Age of New Literacies." In *Engaging Audience: Writing in an Age of New Literacies,* ed. M. Elizabeth Weiser, Brian Fehler, and Angela González, 42–69. Urbana, Ill.: NCTE, 2009.

——. *Singular Texts/Plural Authors: Perspectives on Collaborative Writing.* Carbondale: Southern Illinois University Press, 1990.

MacKendrick, Karmen. *Word Made Skin: Figuring Language at the Surface of the Flesh.* New York: Fordham University Press, 2004.

MacNealy, Mary Sue. *Strategies for Empirical Research in Writing.* New York: Pearson Longman, 1998.

Macy, John A. "Helen Keller as She Really Is." In *The Story of My Life: The Restored Classic,* ed. Roger Shattuck and Dorothy Herrmann, 391–95. New York: Norton, 2003.

Mairs, Nancy. *Plaintext: Deciphering a Woman's Life.* New York: Harper & Row, 1986.

——. *Waist-High in the World: A Life among the Nondisabled.* Boston: Beacon Press, 1997.

Manning, Erin. *The Politics of Touch: Sense, Movement, Sovereignty.* Minneapolis: University of Minnesota Press, 2006.

Marks, Laura U. *Touch: Sensuous Theory and Multisensory Media.* Minneapolis: University of Minnesota Press, 2002.

Martin, Alain, and Oliver Primavesi. *L'Empédocle de Strasbourg.* Berlin: Walter de Gruyter, 1999.

McComiskey, Bruce. *Gorgias and the New Sophistic Rhetoric.* Carbondale: Southern Illinois University Press, 2002.

——."Visual Rhetoric and the New Public Discourse." *JAC: A Journal of Rhetoric, Culture, & Politics* 24.1 (2004): 187–206.

McKeon, Richard. "Creativity and the Commonplace." *Philosophy & Rhetoric* 6.4 (1973): 199–210.

McRuer, Robert. "Composing Bodies; or, De-Composition: Queer Theory, Disability Studies, and Alternative Corporealities." *JAC: A Journal of Rhetoric, Culture, & Politics* 24.1 (2004): 47–77.

——. *Cultural Signs of Queerness and Disability.* New York: New York University Press, 2006.

Mercer, Geof. "Emancipatory Disability Research." In *Disability Studies Today*, ed. C. Barnes, M. Oliver, and L. Barton, 228–49. Malden, Mass.: Polity Press/Blackwell, 2002.

Merleau-Ponty, Maurice. *The Visible and the Invisible.* Trans. Alphonso Lingis. Ed. Claude Lefort. Evanston, Ill.: Northwestern University Press, 1968.

Micciche, Laura R. *Doing Emotion: Rhetoric, Writing, Teaching.* New York: Boynton/Cook Heinemann, 2007.

Miller, Bernard A. "Heidegger and the Gorgian Kairos." In *Visions of Rhetoric: History, Theory, and Criticism*, ed. C. Kneupper, 169–84. Arlington, Va.: Rhetoric Society of America, 1987.

Miller, Carolyn R. "Foreword." In *Rhetoric and Kairos*, ed. Phillip Sipiora and James S. Baumlin, xi–xiii. Albany: State University of New York Press, 2002.

——. "Opportunity, Opportunism, and Progress: *Kairos* in the Rhetoric of Technology." *Argumentation* 8.1 (1994): 81–96.

Milton, John. "Samson Agonistes." In *John Milton: Paradise Lost, Paradise Regained, and Samson Agonistes*, 351–95. Garden City, N.Y.: Doubleday & Company, 1969.

"Mindstorm." Janssen. http://www.janssen.com/mindstorm_video.html (accessed January 30, 2013).

Mitchell, David T., and Sharon L. Snyder. *The Body and Physical Difference: Discourses of Disability.* Ann Arbor: University of Michigan Press, 1997.

Mollow, Anna. "Identity Politics and Disability Studies: A Critique of Recent Theory." *Michigan Quarterly Review* 43.2 (2004): 269–96.

Montagu, Ashley. *Touching: The Human Significance of the Skin.* New York: Harper & Row, 1971.

Mossman, Mark. "Visible Disability in the College Classroom." *College English* 64.6 (2002): 645–59.

"Motorola Droid 2." Motorola. http://www.youtube.com/watch?v=q8bSLMcerCc (accessed January 30, 2013).

Mukhopadhyay, Tito Rajarshi. "Questions and Answers." In *Autism and the Myth of the Person Alone*, ed. Douglas Biklen, 117–43. New York: New York University Press, 2005.

Mullins, Aimee. Aimeemullins.com. "Themes." http://www.aimeemullins.com/about.php (accessed January 29, 2013).

Murderball. Dir. Dana Adam Shapiro and Henry Alex Rubin. Velocity/Thinkfilm, 2005.

Nancy, Jean-Luc. *Corpus: Perspectives in Philosophy.* Trans. Richard A. Rand. New York: Fordham University Press, 2008.

Nelson, Ted. *The Home Computer Revolution.* Self-published, 1977.

Nielsen, Kim E. *Beyond the Miracle Worker: The Remarkable Life of Anne Sullivan Macy and Her Extraordinary Friendship with Helen Keller.* Boston: Beacon Press, 2009.

Nussbaum, Martha C. *Frontiers of Justice: Disability, Nationality, Species Membership.* Cambridge, Mass.: Harvard University Press, Belknap Press, 2006.

Odell, Lee, and Karen McGrane. "Bridging the Gap: Integrating Visual and Verbal Rhetoric." In *Inventing a Discipline: Rhetoric Scholarship in Honor of Richard E. Young,* ed. Maureen Daly Goggin, 207–36. Urbana, Ill.: National Council of Teachers of English, 2000.

Ogden, Charles K., and I. A. Richards. *The Meaning of Meaning: A Study on the Influence of Language upon Thought and of the Science of Symbolism.* New York: Harcourt, Brace & World, 1946.

Olkin, Rhoda. *Women with Physical Disabilities Who Want to Leave Their Partners: A Feminist and Disability-Affirmative Perspective.* San Francisco: California School of Professional Psychology and Through the Looking Glass Co., 2009.

Olson, Kay. "504 Sit-in Anniversary." http://thegimpparade.blogspot.com/2007/04/504-sit-in-anniversary.html (accessed January 5, 2013).

Onians, Richard Broxton. *The Origins of European Thought: About the Body, the Mind, the Soul, the World, Time and Fate.* Cambridge: Cambridge University Press, 1951.

Paley, F. A. *The Epics of Hesiod.* London: Whittaker, 1883.

Park, Melissa M. "Beyond Calculus: Apple-apple-apple-ike and Other Embodied Pleasures for a Child Diagnosed with Autism in a Sensory Integration Clinic." *Disability Studies Quarterly* 30.1 (2010). http://dsq-sds.org/article/view/1066/1232 (accessed January 30, 2013).

Parr, Ben. "How the iPad Has Changed One 99-Year-Old Woman's Life." *Mashable.* http://mashable.com/2010/04/23/ipad-99-year-old-woman/ (accessed January 29, 2013).

Paterson, Mark. *The Senses of Touch: Haptics, Affects and Technologies.* Oxford: Berg, 2007.

Patterson, Randi, and Gail Corning. "Researching the Body: An Annotated Bibliography for Rhetoric." *Rhetoric Society Quarterly* 27.3 (1997): 5–30.

Perl, Sondra. *Felt Sense: Writing with the Body.* Portsmouth, N.H.: Boynton/Cook Heinemann, 2004.

Phelps, Louise Wetherbee. *Composition as Human Science: Contributions to the Self-Understanding of a Discipline.* Oxford: Oxford University Press, 1988.

Plato. *Gorgias.* Trans. W. R. M. Lamb. Cambridge, Mass.: Harvard University Press, 1925.

——. *Laws.* Trans. R. G. Bury. Cambridge, Mass.: Harvard University Press, 1926.

——. *Meno.* Trans. W. R. M. Lamb. Cambridge, Mass.: Harvard University Press, 1924.

——. *Phaedrus.* Trans. Harold North Fowler. Cambridge, Mass.: Harvard University Press, 1914.

Polansky, Ronald. *Aristotle's De Anima: A Critical Commentary.* Cambridge: Cambridge University Press, 2008.

Posidippus. "On a Statue of Time by Lysippus." In *The Greek Anthology,* vol. 5, trans. W. R. Paton, 325. Cambridge, Mass.: Harvard University Press, 1918.

Poulakos, John. *Sophistical Rhetoric in Classical Greece.* Columbia: University of South Carolina Press, 1995.

——. "Toward a Sophistic Definition of Rhetoric." *Philosophy & Rhetoric* 16.1 (1983): 35–48.

Poulakos, Takis. *Speaking for the Polis: Isocrates' Rhetorical Education.* Columbia: University of South Carolina Press, 1997.

Powell, Malea. "All Our Relations: Contested Space, Contested Knowledge." http://www.ncte.org/library/NCTEFiles/Groups/CCCC/Convention/2011/4C_CallFor_2011b.pdf (accessed January 30, 2013).

The Power of 504. The Disability Rights Education and Defense Fund. 1977. Film.

Pratt, Mary Louise. "Arts of the Contact Zone." *Profession,* 1991, 33–40.

Prelli, Lawrence J., ed. *Rhetorics of Display.* Columbia: University of South Carolina Press, 2006.

Prendergast, Catherine. "And Now, a Necessarily Pathetic Response: A Response to Susan Schweik." *American Literary History* 20.1–2 (2008): 238–44.

——. "On the Rhetorics of Mental Disability." In *Embodied Rhetorics,* ed. James C. Wilson and Cynthia Lewiecki-Wilson, 45–60. Carbondale: Southern Illinois University Press, 2001.

——. "The Unexceptional Schizophrenic: A Post-Postmodern Introduction." *Journal of Literary Disability* 2.1 (2008): 55–62.

Price, Margaret. "Accessing Disability: A Nondisabled Student Works the Hyphen." *College Composition and Communication* 59.1 (2007): 53–76.

——. "Disability Studies Methodology." In *Practicing Research in Writing Studies: Reflections on Ethically Responsible Research,* ed. Pamela Takayoshi and Katrina M. Powell, 159–186. New York: Hampton Press, 2002.

——. "'Her Pronouns Wax and Wane': Psychosocial Disability, Autobiography, and Counter-Diagnosis." *Journal of Literary & Cultural Disability Studies* 3.1 (2009): 11–33.

——. *Mad at School: Rhetorics of Mental Disability and Academic Life.* Ann Arbor: University of Michigan Press, 2011.

——. "Writing from Normal: Critical Thinking and Disability in the Composition Classroom." In *Disability and the Teaching of Writing: A Critical Sourcebook,* eds. Cynthia Lewiecki-Wilson and Brenda Jo Brueggemann, 56–73. Boston: Bedford/St. Martin's, 2008.

Price, Janet, and Margrit Shildrick. "Touch, Ethics and Disability." In *Disability/Postmodernity: Embodying Disability Theory,* ed. Mairian Corker and Tom Shakespeare, 62–75. New York: Continuum Press, 2002.

Prince-Hughes, Dawn. *Songs of the Gorilla Nation: My Journey through Autism.* New York: Harmony Books, 2004.

——, ed. *Aquamarine Blue 5: Personal Stories of College Students with Autism.* Athens, Ohio: Swallow Press, 2002.

"Prisoners of Silence." *Frontline.* PBS. October 19, 1993. http://www.youtube.com/watch?v=HXw8Ksvyt5Y (accessed March 22, 2013).

Quandahl, Ellen. "A Feeling for Aristotle: Emotion in the Sphere of Ethics." In *A Way to Move: Rhetorics of Emotion & Composition Studies,* ed. Dale Jacobs and Laura Micciche, 11–22. Portsmouth, N.H.: Boynton/Cook Heinemann, 2003.

Quintilian. *Institutes of Oratory.* Vol. 2. Trans. John Selby Watson. London: George Bell & Sons, 1892.

Ratcliffe, Krista. *Rhetorical Listening: Identification, Gender, Whiteness.* Carbondale: Southern Illinois University Press, 2005.

Rehabtool.com. "Forum." http://rehabtool.com/forum/ (accessed January 30, 2013).

Reynolds, Nedra. *Geographies of Writing: Inhabiting Places and Encountering Difference.* Carbondale: Southern Illinois University Press, 2004.

Rheingold, Howard. *Virtual Reality.* New York: Simon and Schuster, 1991.

Rice, Jenny Edbauer. "The New 'New': Making the Case for Critical Affect Studies." *Quarterly Journal of Speech* 94.2 (2008): 200–212.

Richardson, Elaine. *African American Literacies.* New York: Routledge, 2003.

Ritchie, Joy, and Kate Ronald, eds. *Available Means: An Anthology of Women's Rhetoric(s).* Pittsburgh: University of Pittsburgh Press, 2001.

Ross, Stephen David. *The Gift of Touch: Embodying the Good.* Albany: State University of New York Press, 1998.

Royster, Jacqueline Jones. *Traces of a Stream: Literacy and Social Change among African American Women.* Pittsburgh: University of Pittsburgh Press, 2000.

——. "When the First Voice You Hear Is Not Your Own." *College Composition and Communication* 47.1 (1996): 29–40.

Rubin, Sue. "A Conversation with Leo Kanner." In *Autism and the Myth of the Person Alone,* ed. Douglas Biklen, 82–109. New York: New York University Press, 2005.

——. "Facilitated Communication Enables Me to Think, to Learn, to Live." http://sue-rubino.tripod.com/Presentations/Learning%20and%20Living.htm (accessed January 30, 2013).

——. "Facilitated Communication: The Key to Success for a Non-verbal Person with Autism." http://sue-rubino.tripod.com/Presentations/ASA%20Natl%202003.htm (accessed January 30, 2013).

Sacks, Oliver. *An Anthropologist on Mars: Seven Paradoxical Tales.* New York: Vintage, 1996.

Savarese, Emily Thorton, and Ralph James Savarese. "The Superior Half of Speaking: An Introduction." *Disability Studies Quarterly* 30.1 (2010). http://www.dsq-sds.org/article/view/1062/1230 (accessed January 30, 2013).

Savarese, Ralph James. *Reasonable People: A Memoir of Autism and Adoption.* New York: Other Press, 2007.

Schweik, Susan. "Lomax's Matrix: Disability, Solidarity, and the Black Power of 504." *Disability Studies Quarterly* 31.1 (2011). http://www.dsq-sds.org/article/view/1371/1539 (accessed January 30, 2013).

——. *The Ugly Laws: Disability in Public.* New York: New York University Press, 2010.

Sedgwick, Eve Kosofsky. *Touching Feeling: Affect, Pedagogy, Performativity.* Durham, N.C.: Duke University Press, 2003.

Selber, Stuart A. *Multiliteracies for a Digital Age.* Carbondale: Southern Illinois University Press, 2004.

Selfe, Cynthia L. "The Movement of Air, the Breath of Meaning: Aurality and Multimodal Composing" *College Composition and Comunication* 60.4 (2009): 616–63.

——, ed. *Multimodal Composition: Resources for Teachers.* Cresskill, N.J.: Hampton Press, 2007.

Selzer, Jack. "Habeas Corpus." In *Rhetorical Bodies,* ed. Jack Selzer and Sharon Crowley, 3–15. Madison: University of Wisconsin Press, 1999.

Selzer, Jack, and Sharon Crowley, eds. *Rhetorical Bodies.* Madison: University of Wisconsin Press, 1999.

Sen, Amartya. *Commodities and Capabilities.* Oxford: Oxford University Press, 1999.

Shakespeare, Tom. *Disability Rights and Wrongs.* New York: Routledge, 2006.

——. "The Social Model of Disability." In *The Disability Studies Reader,* second edition, ed. Lennard Davis, 197–204. New York: Routledge, 2006.

Shakespeare, Tom, and Nicholas Watson. "The Social Model of Disability: An Outdated Ideology?" *Research in Social Science and Disability* 2 (2002): 9–28.

Shakespeare, Tom, Kath Gillespie-Sells, and Dominic Davies. *The Sexual Politics of Disability: Untold Desires.* New York: Cassell, 1996.

Shapiro, Joseph P. *No Pity: People with Disabilities Forging a New Civil Rights Movement.* New York: Times Books Random House, 1993.

Shaw, Barrett. "Introduction." In *The Ragged Edge,* ed. Barrett Shaw, x–xiii. Louisville, Ky.: Avocado Press, 1994.

——, ed. *The Ragged Edge.* Louisville, Ky.: Avocado Press, 1994.

Sheridan, Thomas b. "Human and Machine Haptics in Historical Perspective." http://www.universelle-automation.de/1961_Boston.pdf (accessed January 30, 2013).

"Shiny Autistic." http://wikibin.org/articles/shiny-autistic.html (accessed January 21, 2013).

Siebers, Tobin. *Disability Aesthetics.* Ann Arbor: University of Michigan Press, 2010.

——. *Disability Theory.* Ann Arbor: University of Michigan Press, 2008.

——. "Tender Organs, Narcissism, and Identity Politics." In *Disability Studies: Enabling the Humanities,* ed. Sharon Snyder, Brenda Jo Brueggemann, and Rosemarie Garland-Thomson, 40–55. New York: MLA, 2002.

Silverman, Chloe. *Understanding Autism: Parents, Doctors, and the History of a Disorder.* Princeton, N.J.: Princeton University Press, 2012.

Sinclair, Jim. "Being Autistic Together." *Disability Studies Quarterly* 30.1 (2010). http://dsq-sds.org/article/view/1075/1248 (accessed January 30, 2013).

——. "Why I Dislike 'Person First' Language." http://autismmythbusters.com/general-public/autistic-vs-people-with-autism/jim-sinclair-why-i-dislike-person-first-language/ (accessed January 30, 2012).

Singer, Peter. *Animal Liberation.* 2nd ed. New York: Avon Books, 1990.

Sipiora, Phillip. "The Rhetoric of Time and Timing in the New Testament." In *Rhetoric and Kairos: Essays in History, Theory and Praxis,* ed. Phillip Sipiora and James S. Baumlin, 114–27. Albany: State University of New York Press, 2002.

Sipiora, Phillip, and James S. Baumlin, eds. *Rhetoric and Kairos: Essays in History, Theory and Praxis.* Albany: State University of New York Press, 2002.

Sklar, Howard. "'What the Hell Happened to Maggie?': Stereotype, Sympathy, and Disability in Toni Morrison's 'Recitatif.'" *Journal of Literary & Cultural Disability Studies* 5.2 (2011): 137–54.

Slater, Lauren. *Welcome to My Country.* New York: Random House, 1996.

Smith, Craig R., and Michael J. Hyde. "Rethinking 'the Public': The Role of Emotion in Being-with-Others." *Quarterly Journal of Speech* 77.4 (1991): 446–66.

Smith, Craig R., and Scott Lybarger. "Bitzer's Model Revisited." *Communication Quarterly* 44.2 (1996): 197–213.

Snyder, Sharon L., and David T. Mitchell. *Cultural Locations of Disability.* Chicago: University of Chicago Press, 2006.

Snyder, Sharon L., Brenda Jo Brueggemann, and Rosemarie Garland-Thomson, eds. *Disability Studies: Enabling the Humanities.* New York: MLA, 2002.

The Soloist. Dir. Joe Wright. Dreamworks Video, 2009.

Squier, Susan Merrill. *Liminal Lives: Imagining the Human at the Frontiers of Biomedicine.* Durham, N.C.: Duke University Press, 2004.

——. "Meditation, Disability, and Identity." *Literature and Medicine* 23.1 (2004): 23–45.

Stone, Robert J. "Haptic Feedback: A Brief History from Telepresence to Virtual Reality." In *Haptic Human-Computer Interaction,* proceedings of the First International Workshop Glasgow, U.K., August/September 2000, ed. Stephen Brewster and Roderick Murray-Smith, 1–16. Berlin: Springer, 2001.

Sullivan, Dale L. "Kairos and the Rhetoric of Belief." *Quarterly Journal of Speech* 78.3 (1992): 317–32.

Sullivan, Shannon. *Living across and through Skins: Transactional Bodies, Pragmatism, and Feminism.* Bloomington: Indiana University Press, 2001.

Sutherland, Ivan. "Sketchpad: A Man-Machine Graphical Communication System." In *AFIPS Conference Proceedings,* vol. 28. Washington, D.C.: Thompson, 1963.

——. "The Ultimate Display." In *Proceedings of the International Federation of Information Processing Societies Congress,* vol. 2. Amsterdam: North-Holland, 1965.

Swan, Jim. "Disabilities, Bodies, Voices." In *Disability Studies: Enabling the Humanities,* ed. Sharon L. Snyder, Brenda Jo Brueggemann, and Rosemarie Garland-Thomson, 283–95. New York: MLA, 2002.

——. "Touching Words: Helen Keller, Plagiarism, Authorship." In *The Construction of Authorship: Textual Appropriation in Law and Literature,* ed. Martha Woodmansee and Peter Jaszi, 57–100. Durham, N.C.: Duke University Press, 1994.

Symanski, Christina. "iPad Review (with People with Disabilities in Mind)." http://lifeparalyzed.blogspot.com/2010/04/ipad-review-with-people-with.html (accessed January 29, 2013).

Syverson, Margaret A. *The Wealth of Reality: An Ecology of Composition.* Carbondale: Southern Illinois University Press, 1999.

Thayer, Stephen. "Social Touching." In *Tactual Perception: A Sourcebook,* ed. W. Schiff and E. Foulkes, 263–304. Cambridge: Cambridge University Press, 1982.

Thomas, Carol. *Sociologies of Disability and Illness: Contested Ideas in Disability Studies and Medical Sociology.* London: Palgrave/Macmillan, 2007.

Tollifson, Joan. "Imperfection Is a Beautiful Thing." In *Staring Back: The Disability Experience from the Inside Out,* ed. Kenny Fries, 105–12. New York: Plume Books, 1997.

Trépanier, Simon. "The Structure of Empedocles' Fragment 17." *Essays in Philosophy* 1.1 (2000): 1–16.

Trimbur, John. "Theory of Visual Design." In *Coming of Age: The Advanced Writing Curriculum,* ed. Linda K. Shamoon, Rebecca Moore Howard, Sandra Jamieson, and Robert A. Schwegler, 106–14. Portsmouth, N.H.: Heinemann Boynton/Cook Heinemann, 2000.

Tufte, Edward R. *Visual Explanations: Images and Quantities, Evidence and Narrative.* Cheshire, Conn.: Graphics Press, 1997.

Turing, Alan M. "Computing Machinery and Intelligence." *Mind* 59.236 (1950): 433–60.

Turner, Bryan S. *The Body and Society.* 2nd ed. London: SAGE, 2006.

Vatz, Richard E. "The Myth of the Rhetorical Situation." *Philosophy & Rhetoric* 6.3 (1973): 154–61.

Vidali, Amy. "Performing the Rhetorical Freak Show: Disability, Student Writing, and College Admissions." *College English* 69.6 (2007): 615–41.

———. "Seeing What We Know: Disability and Theories of Metaphor." *Journal of Literary & Cultural Disability Studies* 4.1 (2010): 33–54.

Vital Signs: Crip Culture Talks Back. Dir. David Mitchell and Sharon Snyder. Boston: Fanlight Productions, 1997.

Vitanza, Victor J. *Negation, Subjectivity, and the History of Rhetoric.* Albany: State University of New York Press, 1997.

Wade, Cheryl Marie. "I Am Not One of the." In *The Disability Studies Reader,* second edition, ed. Lennard Davis, 411. New York: Routledge, 2006.

———. "It Ain't Exactly Sexy." In *The Ragged Edge: The Disability Experience from the Pages of the First Fifteen Years of The Disability Rag,* ed. Barrett Shaw, 92–94. Louisville, Ky.: Avocado Press, 1994.

———. "My Hands." In *Vital Signs: Crip Culture Talks Back.* Dir. David Mitchell and Sharon Snyder. Marquette, Mich.: Brace Yourself Productions, 1997.

Walker, Jeffrey. *Rhetoric and Poetics in Antiquity.* Oxford: Oxford University Press, 2000.

Walters, Shannon. "Animal Athena: The Interspecies *Mētis* of Women Writers with Autism." *JAC: A Journal of Rhetoric, Culture, and Politics* 30.3–4 (2010): 683–711.

———. "Autistic *Ethos* at Work: Writing on the Spectrum in Contexts of Professional and Technical Communication." *Disability Studies Quarterly* 31.2 (2011). http://www.dsq-sds.org/article/view/1680/1590 (accessed January 30, 2013).

———. "Toward an Accessible Pedagogy: Dis/ability and Multimodality in the Technical Communication Classroom." *Technical Communication Quarterly* 19.4 (2010): 427–54.

Wang, Z., E. Giannopoulos, M. Slater, A. Peer, and M. Buss. "Handshake: Realistic Human-Robot Interaction in Haptic Enhanced Virtual Reality." *Presence-Teleop Virt* 20.4 (2011): 371–92.

"We Shall Not Be Moved." Audiotape produced for the twentieth anniversary of the 504 demonstration. Available from Disability Rights Education and Defense Fund, Berkeley, Calif.

Welch, Kathleen E. *The Contemporary Reception of Classical Rhetoric: Appropriations of Ancient Discourse.* Hillsdale, N.J.: Lawrence Erlbaum, 1990.

———. *Electric Rhetoric: Classical Rhetoric, Oralism, and a New Literacy.* Cambridge, Mass.: MIT Press, 1999.

Wendell, Susan. *The Rejected Body: Feminist Philosophical Reflections on Disability.* New York: Routledge, 1996.

Werner, Marta L. "Helen Keller and Anne Sullivan: Writing Otherwise." *Interval(le)s* 2.2–3.1 (2008/9): 958–96.

"What Is iPad?" *Apple.* http://www.youtube.com/watch?v=D2BvVcSkNkA (accessed January 30, 2013).

White, Eric Charles. *Kaironomia: On the Will-to-Invent.* Ithaca, N.Y.: Cornell University Press, 1987.

Williams, Raymond. *Marxism and Literature.* Oxford: Oxford University Press, 1977.

Wilson, Anne, and Peter Beresford. “Madness, Distress and Disability: Putting the Record Straight.” In *Disability/Postmodernity: Embodying Disability Theory*, ed. Mairian Corker and Tom Shakespeare, 143–58. London: Continuum Press, 2002.

Wilson, James C. “Making Disability Visible: How Disability Studies Might Transform the Medical and Science Writing Classroom.” *Technical Communication Quarterly* 9.2 (2000): 149–61.

———. *Weather Reports from the Autism Front: A Father's Memoir of His Autistic Son*. Jefferson, N.C.: McFarland, 2008.

Wilson, James C., and Cynthia Lewiecki-Wilson. “Disability, Rhetoric and the Body.” In *Embodied Rhetorics: Disability in Language and Culture*, ed. James C. Wilson and Cynthia Lewiecki-Wilson, 1–24. Carbondale: Southern Illinois University Press, 2001.

———, eds. *Embodied Rhetorics: Disability in Language and Culture*. Carbondale: Southern Illinois University Press, 2001.

Woolley, Benjamin. *Virtual Worlds: A Journey in Hype and Hyperreality*. Oxford: Blackwell, 1992.

Wright, M. R. *Empedocles: The Extant Fragments*. London: Bristol Classical Press, 1995.

Wyschogrod, Edith. “Doing Before Hearing: On the Primacy of Touch.” In *Textes pour Emmanuel Levanis*, ed. F. Laruelle, 179–203. Paris: Jean Laplace, 1980.

Wysocki, Anne Frances. “Seeing the Screen: Research into Visual and Digital Writing Practices.” In *Handbook of Research on Writing*, ed. Charles Bazerman, 599–612. New York: Lawrence Erlbaum, 2008.

Wysocki, Anne Frances, Johndan Johnson-Eilola, Cynthia L. Selfe, and Geoffrey Sirc. *Writing New Media*. Logan: Utah State University Press, 2004.

Yergeau, Melanie. “Aut(hored)ism.” *Computers and Composition Online*. 2009. http://www.bgsu.edu/cconline/dmac/index.html (accessed January 30, 2013).

———. “Circle Wars: Reshaping the Typical Autism Essay.” *Disability Studies Quarterly* 30.1 (2010). http://www.dsq-sds.org/article/view/1063/1222 (accessed January 30, 2013).

Ziegler, Matilda. “Blind iPad Review.” *Matilda Ziegler Magazine for the Blind*. http://www.matildaziegler.com/2010/04/12/blind-ipad-review/ (accessed January 30, 2013).

Zielinski, Siegfried. *Deep Time of the Media: Toward an Archaeology of Hearing and Seeing by Technical Means*. Trans. Gloria Custance. Cambridge, Mass.: MIT Press, 2006.

Zola, Irving Kenneth. “Self, Identity and the Naming Question: Reflections on the Language of Disability.” *Social Science and Medicine* 36.2 (1993): 167–73.

Index

Page references given in *italics* indicate illustrations or material contained in their captions.

About the Author

SHANNON WALTERS is an assistant professor of English at Temple University, where she teaches courses in rhetoric and composition, disability studies, and women's studies. Her work has appeared in *JAC: A Journal of Rhetoric, Culture & Politics; Technical Communication Quarterly; Feminist Media Studies; Disability Studies Quarterly;* and *Journal of Literary and Cultural Disability Studies* as well as *PMLA.*